Explicit *a priori* inequalities with applications to boundary value problems

V G Sigillito
The Johns Hopkins University

Explicit *a priori* inequalities with applications to boundary value problems

Pitman Publishing
LONDON · SAN FRANCISCO · MELBOURNE

PITMAN PUBLISHING LIMITED
39 Parker Street, London WC2B 5PB

PITMAN PUBLISHING CORPORATION
6 Davis Drive, Belmont, California 94002, USA

Associated Companies
Copp Clark Ltd, Toronto · Fearon Publishers Inc, Belmont, California
Pitman Publishing Co. SA (Pty) Ltd, Johannesburg · Pitman Publishing
New Zealand Ltd, Wellington · Pitman Publishing Pty Ltd, Melbourne

AMS Subject Classifications: (main) 35A40
(subsidiary) 65N99, 35J25, 35J40, 35K20

Library of Congress Cataloging in Publication Data

Sigillito, V. G. 1937–
Explicit a priori inequalities with applications to boundary value problems.

(Research notes in mathematics; 13)
1. Boundary value problems. 2. Differential equations, Parabolic. 3. Differential equations, Elliptic.
4. Inequalities (Mathematics) I. Title. II. Series.
QA379.S57 515'.35 77–1199
ISBN 0–273–01022–0

Reproduced and printed by photolithography
in Great Britain at Biddles of Guildford

To Barbara, Robert and Amanda

Contents

1 Introduction

This research note has several objectives. One is to bring together in a single volume a number of explicit a priori inequalities which are useful in computing approximate solutions to boundary value problems which arise in such physical applications as heat and mass transfer, potential theory, fluid dynamics, elasticity, and radiation diffusion. Mathematically this means that we shall be concerned with elliptic boundary value problems and parabolic and pseudoparabolic initial-boundary value problems.

A second objective is to illustrate a method, based on the a priori inequalities, for computing approximate solutions with error bounds for the various boundary value problems. This is accomplished by way of examining a number of numerical examples. An important feature of the method is that the trial functions used in the approximation need not satisfy either the differential equation or the boundary conditions.

Another objective is to indicate the techniques used to derive the inequalities. Thus the interested reader will be able to develop new a priori inequalities for problems of interest if such problems are not covered by the inequalities presented here or by those in the literature. For this reason we give detailed derivations of most of the inequalities. To develop each inequality in its most general form would require many details which do not aid understanding but rather tend to obscure the main ideas. Therefore we treat special cases of each inequality but in such a manner that extensions to more general cases are evident. There is at least one occasion when this approach results in an inequality which is not the best possible one for the specific problem at hand. When this happens we shall also derive the better inequality.

When applied to boundary value problems the usefulness of the *a priori* inequalities lies in the following areas:

1. They provide a practical method of *a posteriori* pointwise error estimation for any sufficiently smooth approximate solution.
2. For linear problems they can be used with a Ritz type procedure to obtain approximate solutions with norm error bounds. These norm error bounds can be combined with other inequalities to give pointwise error bounds.

Although it was the first area which originally stimulated the research in explicit *a priori* inequalities [28], [30], it is the second area of application which we shall emphasize in this volume.

THE "METHOD OF EXPLICIT A PRIORI INEQUALITIES"

For our purposes an explicit *a priori* inequality is one in which a bound on the L_2 integral of an arbitrary sufficiently smooth function u (a term to be made precise for each specific problem) is given in terms of L_2 integrals of u which represent data of a specific boundary value problem. In addition all constants which appear in the inequality must be known explicitly or be computable. Once the appropriate inequality is in hand the method of explicit *a priori* inequalities is quite straightforward.

For instance, if we are interested in the Dirichlet problem

$$\begin{aligned} \Delta u = \frac{\partial^2 u}{\partial x_1^2} + \frac{\partial u^2}{\partial x_2^2} &= f(x_1, x_2) \quad \text{in} \quad B , \\ u = g(x_1, x_2) & \qquad\qquad \text{on} \ \partial B , \end{aligned} \tag{1.1}$$

then an appropriate *a priori* inequality is

$$\int_B w^2 dx \le \alpha_1 \int_B (\Delta w)^2 dx + \alpha_2 \oint_{\partial B} w^2 dS \ , \tag{1.2}$$

where w is an arbitrary $C^2(B)$ function. The constants are explicitly known and are functions of the geometry of the domain B, certain eigenvalues, the coefficients of the differential equation, and possibly, known auxiliary functions.

Having this inequality we are now ready to apply the method. We introduce an approximate solution $u_a = \sum_{k=1}^{n} a_k \psi_k$ as a linear combination of the trial functions ψ_k, which, because of the a priori nature of (1.2) need not satisfy either the boundary conditions or the differential equation but only be C^2. If we denote the solution of (1.1) by u, then $u - u_a \in C^2(B)$ and we can substitute this function into (1.2) to obtain

$$\begin{aligned} \int_B (u-u_a)^2 dx &\le \alpha_1 \int_B (\Delta u - \Delta u_a)^2 dx + \alpha_2 \oint_{\partial B} (u-u_a)^2 dS \\ &= \alpha_1 \int_B (f - \sum_{k=1}^{n} a_k \Delta\psi_k)^2 dx + \alpha_2 \oint_{\partial B} (g - \sum_{k=1}^{n} a_k \psi_k)^2 dS \ , \end{aligned} \tag{1.3}$$

giving immediately a norm bound on the error of the approximation. Further, since f,g and ψ_k, $k=1,\cdots,n$ are known functions, the right hand side is a function only of the coefficients $a_1,\cdots,a_n$. Denote this function by E and minimize it with respect to the variables $a_1,\cdots,a_n$. This leads to the system of n equations in the n unknowns $a_1,\cdots,a_n$:

$$\begin{aligned} &\sum_{k=1}^{n} \{\alpha_1 \int_B \Delta\psi_k \Delta\psi_i dx + \alpha_2 \oint_{\partial B} \psi_k \psi_i dS\} \, a_k \\ &= \alpha_1 \int_B f \, \Delta\psi_i dx + \alpha_2 \oint_{\partial B} g \, \psi_i dS \ , \qquad i=1,\cdots,n \ . \end{aligned} \tag{1.4}$$

We denote the solution of this system by $a_1^*,\cdots,a_n^*$ and thus obtain the approximate solution $u_a = \sum_{k=1}^{n} a_k^* \psi_k$ which is the best approximation to u in the sense that the bound $E(a_1^*, \cdots, a_n^*)$ on $\int_B (u-u_a)^2 dx$ is a minimum.

Usually more can be said than this. If we denote by Γ either a fundamental solution of Δ or a parametrix then we can obtain a pointwise bound on $|u(P)-u_a(P)|$ of the form

$$|u(P)-u_a(P)|^2 \leq \{\int_{K_1} (\Delta(\sigma\Gamma_P))^2 dx + \int_{K_1} (\sigma\Gamma_P)^2 dx\} \cdot E(a_1^*, a_2^*, \cdots, a_n) \quad (1.5)$$

where $P \in B$, K_1 is a region containing P and contained in B such that the distance $d(K_1, B) > 0$ and σ is a function which is identically 1 in a subregion K_2 of K_1, and belongs to $C^2(B)$.

A RELATIONSHIP TO LEAST SQUARES

Our method bears some relationship to the method of least squares. For instance, obtaining a least squares approximation to (1.1) involves minimizing the functional

$$I = w_1 \int_B (f-\Delta u_a)^2 dx + w_2 \oint_{\partial B} (g-u_a)^2 dS$$

for $u_a = \sum a_k \psi_k$, with respect to the a_i, exactly as is done in the method of <u>a priori</u> inequalities. This procedure leads to a system of equations similar to the system (1.4). There are important differences however. Thus, contrary to our method, there is no <u>explicit</u> relationship between I and either the norm or pointwise difference $u-u_a$. A separate analysis is needed to show that as $I \to 0$ the same is true for $u-u_a$, or some functional of $u-u_a$, for the selected trial functions ψ_k. Also, with least squares, one has the additional task of selecting good values for the weights w_1, w_2, whereas in our method the corresponding constants α_1, α_2 are closely related to the problem at hand and in some cases can be shown to be optimal.

OUTLINE OF CHAPTERS

This work is divided into chapters as follows. In Chapter 2 we introduce notation and develop some important identities. The next chapter is devoted to eigenvalues which are important in the development of the *a priori* inequalities. Chapters 4 through 7 develop the *a priori* inequalities applicable to second order elliptic and parabolic problems, third order pseudo-parabolic problems and fourth order elliptic problems. In Chapter 8 the pointwise bounds are discussed. Chapter 9 presents recent work on the use of *a priori* inequalities in eigenvalue estimation. The numerical examples are given in Chapter 10.

2 Notation and some important identities and inequalities

In this chapter we present the notation which will be used in the remainder of the book. We then summarize some well-known identities and inequalities that are used over and over again in the development of the *a priori* inequalities. Finally we derive three identities which play a central role in the chapters that follow.

NOTATION

Although most of the numerical examples will be carried out in two dimensional regions we shall present the inequalities for functions of N variables since this generality is obtained with no additional effort. We shall use the convention that $,_i$ denotes partial differentiation with respect to the variable x_i so that, for instance, $u,_i = \partial u/\partial x_i$. The summation convention is also used so that repeated indices in an expression are to be summed from 1 to N and thus, for example, the Laplacian Δu can be expressed as $u,_{ii}$. In the parabolic cases where a time variable t is present in addition to the spatial variables $x = (x_1, x_2, \cdots, x_N)$, repeated Latin indices are to be summed from 1 to N while repeated Greek indices are to be summed from 1 to N+1, the N+1-st variable being time.

The elliptic problems will be defined in bounded regions B of N-dimensional Euclidean space whose boundary we denote by ∂B. No other restrictions are placed on B except that ∂B is smooth enough so that the divergence theorem applies in B (requiring ∂B to be piecewise C^1 is sufficient). On ∂B we denote the outward pointing unit normal by $n = (n_1, n_2, \cdots, n_N)$ and the normal derivative of u by $\partial u/\partial n = u,_i n_i$.

The parabolic problems will be defined in a region $D = B \times (0,T]$, $T < \infty$, i.e., a space-time right cylinder with base B. Then we define $B_\tau \equiv D \cap \{t=\tau\}$, the sides $S \equiv \partial B \times (0,T]$ and the top $B_T = \bar{D} \cap \{t=T\}$. The so-called parabolic boundary of D is thus $\bar{B} + S$.

SOME IMPORTANT IDENTITIES AND INEQUALITIES

For ease of reference we now give a number of well-known theorems, identities and inequalities which will be used in the development of the a priori inequalities.

(A) The divergence theorem (integration by parts). Let $f_i = f_i(x)$, $i=1,\cdots,N$ and $g_\alpha = g_\alpha(x,t)$, $\alpha=1,\cdots,N+1$ denote smooth vector fields, then

$$\int_B f_{i,i}\,dx = \oint_{\partial B} f_i n_i\,dS \quad ,$$

$$\int_D g_{\alpha,\alpha}\,dV = \int_S g_i n_i\,d\sigma + \int_{B_T} g_{N+1}\,dx - \int_B g_{N+1}\,dx \ ,$$

and in a form we shall frequently use

$$\int_B h\, f_{i,i}\,dx = \oint_{\partial B} h\, f_i n_i\,dS - \int_B h,_i f_i\,dx \quad .$$

(B) Green's first identity:

$$\int_B u\,\Delta v\;dx + \int_B u,_i v,_i\,dx = \oint_{\partial B} u\,\frac{\partial v}{\partial n}\,dS \ .$$

(C) Green's second identity:

$$\int_B (u\,\Delta v - v\,\Delta u)\,dx = \oint_{\partial B} \left(u\,\frac{\partial v}{\partial n} - v\,\frac{\partial u}{\partial n}\right)\,dS \ .$$

(D) The arithmetic-geometric mean inequality (hereafter referred to as the a-g inequality) with weight $\alpha > 0$:

$$2ab \le \frac{a^2}{\alpha} + \alpha b^2 \ .$$

This inequality will most frequently be used in the integral form

$$2 \int_B u\, v\, dx \le \int_B \alpha\, u^2 dx + \int_B \alpha^{-1} v^2 dx \ .$$

(E) The Schwarz inequality:

$$\int_B u\, v\, dx \le \left(\int_B u^2 dx\right)^{\frac{1}{2}} \left(\int_B v^2 dx\right)^{\frac{1}{2}}$$

and its vector form

$$\int_B f_i g_i dx \le \left(\int_B \sum_{i=1}^{N} f_i^2\ dx\right)^{\frac{1}{2}} \left(\int_B \sum_{i=1}^{N} g_i^2\ dx\right)^{\frac{1}{2}} \ .$$

ADDITIONAL IDENTITIES

The following three identities are less well-known but are crucial in the development of the inequalities which apply to the second order elliptic and parabolic equations and the third order pseudoparabolic equations. The first identity is due to Payne and Weinberger [30] and contains the Rellich identity [37] as a special case. The other two identities are due to the author [39], [44].

Let f^i denote the i-th component of a piecewise $C^1(B)$ vector field and u a sufficiently smooth function (say piecewise $C^2(B)$). Then starting with $\int_B f^i u_{,i}\, \Delta u\, dx$, a straightforward application of Green's first identity and integration by parts results in the identity

$$\oint_{\partial B} \{f^i n_i u_{,j} u_{,j} - 2 f^i u_{,i} u_{,j} n_j\} dS = - 2 \int_B f^i u_{,i}\, \Delta u\, dx + \int_B \{f^i_{,i} u_{,j} u_{,j} - 2 f^i_{,j} u_{,i} u_{,j}\} dx \ . \tag{2.1}$$

If $f^i = x_i$ this is just the Rellich identity. We now use the fact that

$$\mathcal{S}_i = u,_i - n_i \frac{\partial u}{\partial n} \tag{2.2}$$

is orthogonal to n_i and hence is a tangent vector. If we normalize

$$s_i = \mathcal{S}_i / (\mathcal{S}_j \mathcal{S}_j)^{\frac{1}{2}} \tag{2.3}$$

and introduce the notation

$$\frac{\partial u}{\partial s} = u,_i s_i \tag{2.4}$$

we obtain from (2.2) and (2.3) that

$$u,_i u,_i = (\partial u/\partial s)^2 + (\partial u/\partial n)^2 \quad . \tag{2.5}$$

Using (2.2)-(2.5) in (2.1) gives the first identity

$$\oint_{\partial B} [f^i n_i \{(\partial u/\partial s)^2 - (\partial u/\partial n)^2\} - 2f^i s_i (\partial u/\partial n)(\partial u/\partial s)]dS$$
$$= -2\int_B f^i u,_i \, \Delta u \, dx + \int_B \{f^i,_i u,_j u,_j - 2f^i,_j u,_i u,_j\}dx \, . \tag{2.6}$$

Analogous identities which will be useful in the derivation of the *a priori* inequalities applicable to parabolic and pseudoparabolic problems are derived in a similar manner starting with

$$\int_S \{f^\alpha n_\alpha u,_j u,_j - 2f^\alpha u,_\alpha u,_j n_j\}d\sigma = -2\int_D f^\alpha u,_\alpha \begin{Bmatrix} L^* u - u_t \\ L^* u + \Delta u_t - u_t \end{Bmatrix} dV$$

$$+ \int_D \{f^\alpha,_\alpha u,_j u,_j - 2f^\alpha,_j u,_\alpha u,_j\}dV$$

where $u_t \equiv \partial u/\partial t$ (we do not use the comma notation for time derivatives),

$L^*u = \Delta u + u_t$ and $\mathcal{L}^* = \Delta(u-u_t) + u_t$. These identities are

$$\int_S [f^i n_i\{(\partial u/\partial s)^2 - (\partial u/\partial n)^2\} - 2f^i s_i(\partial u/\partial n)(\partial u/\partial s)]d\sigma$$

$$= -2\int_D f^\alpha u,_\alpha L^*u\, dV + \int_D \{f^\alpha,_\alpha u,_i u,_i - 2f^j,_i u,_i u,_j]dV$$

$$- 2\int_D f^{N+1},_i u_t u,_i dV + 2\int_D f^\alpha u,_\alpha u_t dV + 2\int_S f^{N+1} u_t\, \partial u/\partial n\, d\sigma \tag{2.7}$$

$$- \int_{B+B_T} f^\alpha n_\alpha u,_i u,_i dx .$$

and

$$\int_S [f^i n_i\{(\partial u/\partial n)^2 - (\partial u/\partial n)^2\} - 2f^i s_i(\partial u/\partial n)(\partial u/\partial s)]d\sigma$$

$$= -2\int_D f^\alpha u,_\alpha \mathcal{L}^* u\, dV + \int_D \{f^\alpha,_\alpha u,_i u,_i - 2f^j,_i u,_i u,_i\}dV$$

$$- 2\int_D f^{N+1},_i u_t u,_i dV + 2\int_D f^\alpha u,_\alpha (u_t - \Delta u_t) dV + 2\int_S f^{N+1} u_t\, \partial u/\partial n\, d\sigma \tag{2.8}$$

$$- \int_{B+B_T} f^\alpha n_\alpha u,_i u,_i dx .$$

3 Eigenvalue problems

The eigenvalue inequalities discussed in this chapter have wide application to the *a priori* inequalities. In fact it can be shown that the lowest non-zero eigenvalues of certain eigenvalue problems are the optimal constants in the *a priori* inequalities. Since exact values of the eigenvalues are known only for special regions, the problem of finding bounds for these eigen-values is of importance. We devote the last section of the chapter to this topic.

EIGENVALUE INEQUALITIES

The inequalities we are interested in here represent many of the plausible inequalities among the expressions

$$\int_B u^2\,dx, \qquad \oint_{\partial B} u^2\,dS, \qquad \int_B u,_i u,_i\,dx, \qquad \oint_{\partial B} \left(\frac{\partial u}{\partial n}\right)^2 dS, \qquad \int_B (\Delta u)^2\,dx,$$

for functions u satisfying various smoothness and auxiliary conditions. The inequalities all have the common property that they arise from variational characterizations of various eigenvalues. They are

$$\int_B u^2\,dx \le q^{-1} \oint_{\partial B} u^2\,dS, \qquad \Delta u = 0 \quad \text{on} \quad B; \tag{3.1}$$

$$\int_B u^2\,dx \le \lambda^{-1} \int_B u,_i u,_i\,dx, \qquad u = 0 \quad \text{on} \quad \partial B; \tag{3.2}$$

$$\int_B u^2\,dx \le \mu^{-1} \int_B u,_i u,_i\,dx, \qquad \int_B u\,dx = 0; \tag{3.3}$$

$$\int_B u^2\,dx \le \xi^{-1} \oint_{\partial B} \left(\frac{\partial u}{\partial n}\right)^2 dS, \qquad \Delta u = 0 \quad \text{on} \quad B, \qquad \int_B u\,dx = 0; \tag{3.4}$$

$$\int_B u^2 dx \le \Omega^{-1} \int_B (\Delta u)^2 dx, \qquad u = \frac{\partial u}{\partial n} = 0 \quad \text{on} \quad \partial B; \tag{3.5}$$

$$\int_B u^2 dx \le \lambda^{-2} \int_B (\Delta u)^2 dx, \qquad u = 0 \quad \text{on} \quad \partial B; \tag{3.6}$$

$$\int_B u^2 dx \le \mu^{-2} \int_B (\Delta u)^2 dx, \qquad \frac{\partial u}{\partial n} = 0 \quad \text{on} \quad \partial B, \qquad \int_B u\, dx = 0; \tag{3.7}$$

$$\oint_{\partial B} u^2 dS \le p^{-1} \int_B u,_i u,_i dx, \qquad \oint_{\partial B} u\, dS = 0; \tag{3.8}$$

$$\oint_{\partial B} u^2 dS \le p^{-2} \oint_{\partial B} \left(\frac{\partial u}{\partial n}\right)^2 dS, \qquad \Delta u = 0 \quad \text{on} \quad B, \qquad \oint_{\partial B} u\, dS = 0; \tag{3.9}$$

$$\oint_{\partial B} u^2 dS \le \xi^{-1} \int_B (\Delta u)^2 dx, \qquad \frac{\partial u}{\partial n} = 0 \quad \text{on} \quad \partial B, \qquad \oint_{\partial B} u\, dS = 0; \tag{3.10}$$

$$\int_B u,_i u,_i dx \le p^{-1} \oint_{\partial B} \left(\frac{\partial u}{\partial n}\right)^2 dS, \qquad \Delta u = 0 \quad \text{on} \quad B; \tag{3.11}$$

$$\int_B u,_i u,_i dx \le \Lambda^{-1} \int_B (\Delta u)^2 dx, \qquad u = \frac{\partial u}{\partial n} = 0 \quad \text{on} \quad \partial B; \tag{3.12}$$

$$\int_B u,_i u,_i dx \le \lambda^{-1} \int_B (\Delta u)^2 dx, \qquad u = 0 \quad \text{on} \quad \partial B; \tag{3.13}$$

$$\int_B u,_i u,_i dx \le \mu^{-1} \int_B (\Delta u)^2 dx, \qquad \frac{\partial u}{\partial n} = 0 \quad \text{on} \quad \partial B; \tag{3.14}$$

$$\oint_{\partial B} \left(\frac{\partial u}{\partial n}\right)^2 dS \le q^{-1} \int_B (\Delta u)^2 dx, \qquad u = 0 \quad \text{on} \quad \partial B. \tag{3.15}$$

The optimal constants in the above inequalities are the reciprocals of the <u>first</u> <u>non-zero</u> eigenvalues of the following problems:

The fixed membrane

$$\Delta u + \lambda u = 0 \quad \text{on} \quad B, \qquad u = 0 \quad \text{on} \quad \partial B; \tag{3.16}$$

the free membrane

$$\Delta u + \mu u = 0 \quad \text{on} \quad B, \qquad \frac{\partial u}{\partial n} = 0 \quad \text{on} \quad \partial B; \tag{3.17}$$

the clamped plate

$$\Delta^2 u - \Omega u = 0 \quad \text{on} \quad B, \qquad u = \frac{\partial u}{\partial n} = 0 \quad \text{on} \quad \partial B; \tag{3.18}$$

the buckling of the clamped plate

$$\Delta^2 u - \Lambda\, \Delta u = 0 \quad \text{on} \quad B, \qquad u = \frac{\partial u}{\partial n} = 0 \quad \text{on} \quad \partial B; \tag{3.19}$$

and the Stekloff problems,

$$\Delta u = 0 \quad \text{on} \quad B, \qquad \frac{\partial u}{\partial n} = p\, u \quad \text{on} \quad \partial B; \tag{3.20}$$

$$\Delta^2 u = 0 \quad \text{on} \quad B, \qquad \frac{\partial u}{\partial n} = \frac{\partial \Delta u}{\partial n} + \xi\, u = 0 \quad \text{on} \quad \partial B; \tag{3.21}$$

$$\Delta^2 u = 0 \quad \text{on} \quad B, \qquad u = \Delta u - q\, \frac{\partial u}{\partial n} = 0 \quad \text{on} \quad \partial B. \tag{3.22}$$

We denote the optimal constants in the inequalities (3.1)-(3.15) by λ_1, μ_2, p_2, Ω_1, Λ_1, q_1 and ξ_2 ($\mu_1 = p_1 = \xi_1 = 0$).

That the eigenvalue problems (3.16)-(3.22) do indeed furnish the optimal constants as indicated in (3.1)-(3.15) is easily shown. From (3.2) we see that λ_1 is the minimum of the Rayleigh quotient

$$\int_B u_{,i} u_{,i}\, dx \Big/ \int_B u^2\, dx$$

over all admissible functions u satisfying $u = 0$ on ∂B. The Euler equation obtained by taking the first variation of this Rayleigh quotient is (3.16). Similarly, (3.17), (3.18), (3.19), (3.20), (3.21), and (3.22) are the Euler equations for the Rayleigh quotients for (3.3), (3.5), (3.12), (3.8), (3.10), and (3.15), respectively.

We use Fichera's Principle of Duality [11] to show the equivalence of (3.1) and (3.15), (3.2) and (3.13), (3.3) and (3.14), (3.4) and (3.10), (3.8) and (3.11). For example, to show that (3.4) is equivalent to (3.10), suppose

(3.4) holds and we wish to establish (3.10). Given u satisfying $\partial u/\partial n = 0$ on ∂B, $\oint_{\partial B} u\, dS = 0$, define v by

$$\Delta v = 0 \quad \text{on} \quad B, \qquad \frac{\partial v}{\partial n} = u \quad \text{on} \quad \partial B, \qquad \int_B v\, dx = 0.$$

Then, by Green's second identity, Schwarz's inequality, and (3.4)

$$\oint_{\partial B} u^2 dS = \oint_{\partial B} u \frac{\partial v}{\partial n} dS = -\int_B v\, \Delta u\, dx \le \left(\int_B v^2 dx\right)^{\frac{1}{2}} \left(\int_B (\Delta u)^2 dx\right)^{\frac{1}{2}}$$

$$\le \left(\xi^{-1} \oint_{\partial B} \left(\frac{\partial v}{\partial n}\right)^2 dS\right)^{\frac{1}{2}} \left(\int_B (\Delta u)^2 dx\right)^{\frac{1}{2}} = \left(\oint_{\partial B} u^2 dS\right)^{\frac{1}{2}} \left(\xi^{-1} \int_B (\Delta u)^2 dx\right)^{\frac{1}{2}},$$

implying (3.10). Conversely, suppose (3.10) holds and we are given u satisfying $\Delta u = 0$ on B, $\oint_{\partial B} u\, dS = 0$. Define v by

$$\Delta v = u \quad \text{on} \quad B, \qquad \frac{\partial v}{\partial n} = 0 \quad \text{on} \quad \partial B, \qquad \oint_{\partial B} v\, dS = 0$$

Then, by Green's second identity, Schwarz's inequality, and (3.10),

$$\int_B u^2 dx = \int_B u\, \Delta u\, dx = -\oint_{\partial B} v \frac{\partial u}{\partial n} dS \le \left(\oint_{\partial B} v^2 dS\right)^{\frac{1}{2}} \left(\oint_{\partial B} \left(\frac{\partial u}{\partial n}\right)^2 dS\right)^{\frac{1}{2}}$$

$$\le \left(\xi^{-1} \int_B (\Delta v)^2 dx\right)^{\frac{1}{2}} \left(\oint_{\partial B} \left(\frac{\partial u}{\partial n}\right)^2 dS\right)^{\frac{1}{2}} = \left(\int_B u^2 dx\right)^{\frac{1}{2}} \left(\xi^{-1} \oint_{\partial B} \left(\frac{\partial u}{\partial n}\right)^2 dS\right)^{\frac{1}{2}},$$

implying (3.4).

That λ_1^{-1} and μ_2^{-2} are the optimal constants in (3.6) and (3.7) follows from the Euler equations

$$\Delta^2 u - \lambda^2 u = 0 \quad \text{on} \quad B, \qquad u = \Delta u = 0 \quad \text{on} \quad \partial B,$$

$$\Delta^2 u - \mu^2 u = 0 \quad \text{on} \quad B, \qquad \frac{\partial u}{\partial n} = \frac{\partial \Delta u}{\partial n} = 0 \quad \text{on} \quad \partial B,$$

which are satisfied by the eigenfunctions of (3.2) and (3.3) respectively, and these form complete orthonormal sets for their respective classes of admissible functions.

Finally, combining (3.8) and (3.11) yields (3.9) and equality holds for u_2, the eigenfunction of (3.20) associated with p_2, so the optimal constant is p_2^{-2}.

BOUNDS FOR THE EIGENVALUES IN SPECIFIC REGIONS

Precise values of λ_1, μ_2, p_2, ξ_2, q_1, Ω_1, and Λ_1 are known only for certain simple regions. For instance for a disc B of radius R we have

$$\lambda_1 = j^2/R^2 \approx 5.783/R^2 \; ; \tag{3.23}$$

$$\mu_2 = y^2/R^2 \approx 3.39/R^2 \; ; \tag{3.24}$$

$$\Omega_1 = \ell^4/R^4 \approx 103.6/R^4 \; ; \tag{3.25}$$

$$\Lambda_1 = z^2/R^2 \approx 11.446/R^2 \; ; \tag{3.26}$$

$$p_2 = 1/R \; ; \tag{3.27}$$

$$\xi_2 = 5/R^3 \; ; \tag{3.28}$$

$$q_1 = 2/R \; . \tag{3.29}$$

In the above, j is the first zero of the zero order Bessel function of the first kind, J_0, z is the first zero of J_0', y is the first zero of J_1' and ℓ is the first zero of $J_0 I_0' - J_0' I_0$, where $I_0(x) = J_0(ix)$, $i = \sqrt{-1}$. Exact values of λ_1, μ_2 and p_2 are also known for rectangular regions.

In the applications of the eigenvalue inequalities to the development of a priori inequalities, we do not need to know the precise values of the above eigenvalues but only lower bounds. We indicate some useful results along these lines.

(A) The Faber-Krahn inequality [10] which states that the first eigenvalue of problem (3.16) on B is not smaller than that for the sphere whose N-volume is the same as that of B. The mathematical statement is

$$\lambda_1 \geq \left(\frac{\omega_N}{N\, V_N}\right) j^2_{(N-2)/2} \; . \tag{3.30}$$

Here N denotes the number of dimensions, ω_N the surface area of the N-dimensional unit sphere, V_N the volume of B and $j_{(N-2)/2}$ the first zero of the Bessel function $J_{(N-2)/2}$.

For two dimensional regions Payne and Weinberger [31] have extended the Faber-Krahn inequality as follows:

If B lies interior to the wedge of angle π/α, i.e., $0 \leq \theta \leq \pi/\alpha$, for any real $\alpha \geq 1$, then

$$\lambda_1 \geq \left\{\frac{\pi}{4} \cdot \frac{K_\alpha^{-1}}{\alpha(\alpha+1)}\right\}^{1/(\alpha+1)} j^2_\alpha \; , \tag{3.31}$$

where

$$K_\alpha = \int_B r^{2\alpha} \sin^2 \alpha\, \theta \; d\theta \; . \tag{3.32}$$

Equality holds if and only if B is a circular sector.

(B) Monotony principles: For regions A and B, $A \supset B$,

$$\lambda_1(A) \leq \lambda_1(B) \; ,$$

$$\Omega_1(A) \leq \Omega_1(B) \; ,$$

$$\Lambda_1(A) \leq \Lambda_1(B) \; .$$

(C) For a convex N-dimensional region B

$$\mu_2 > \pi^2 D^{-2} \tag{3.33}$$

where D is the diameter of B [32].

(D) Consider the case where B is star-shaped with respect to some point which we choose as the origin. Let r(P) denote the distance from this origin to a point $P \in \partial B$ and let h(P) be the distance from the origin to the tangent plane to ∂B at P. Then

$$\mu_2 \geq \frac{N}{2r_M^2\left\{\left(\frac{r_M}{r_m}\right)^{N-1} \cdot \frac{r_M}{h_m} + \frac{2}{N}\right\}} \tag{3.34}$$

and

$$p_2 \geq \frac{1}{r_M}\left\{\left(\frac{r_m}{r_M}\right)^{N-1} \frac{h_m}{r_M}\right\}$$

Here the subscripts M and m denote the maximum and minimum values of the associated quantities. We note that $h = x_i n_i$ where x_i denotes the i-th component of P and n_i denotes the i-th component of n at P. The star-shapedness of B insures that $h_m > 0$. The above results are due to Bramble and Payne [3].

(E) The following inequalities are also useful:

$$q_1 \geq \frac{\lambda_1^{\frac{1}{2}}}{2} \frac{h_m}{r_M} \quad , \quad \text{B star-shaped,} \tag{3.36}$$

$$q_1 \geq p_2 + (1/\rho)_m \quad , \quad \rho \text{ denotes the curvature of } \partial B,, \tag{3.37}$$

$$\Lambda_1 \geq \lambda_2 \quad , \quad \text{equality holds if and only if B is the N-sphere,} \tag{3.38}$$

$$\Omega_1 \geq \Lambda_1 \lambda_1 \quad , \tag{3.39}$$

$$\Omega_1 > \lambda_1 \lambda_2 \ . \tag{3.40}$$

Pertinent references for these results are [18] and [33]. For other useful bounds see [17], [36].

4 *A priori* inequalities I – Second order elliptic applications

We now develop explicit a priori inequalities which have applications to the three classical boundary value problems for second order elliptic equations: the Dirichlet, Neumann and Robin problems. Also included is an a priori inequality which has applications to the first boundary value problem in the equations of elasticity.

The inequalities are given in terms of the Laplacian but the derivation given can be carried over to the general self-adjoint operator $Lu = (a_{ij}u,_i),_j$ where $a_{ij} = a_{ji} = a_{ij}(x)$ is a symmetric matrix such that $a_o\xi_i\xi_i \leq a_{ij}\xi_i\xi_j \leq a_1\xi_i\xi_i$, for all $x \in B$ and all real vectors $\xi = (\xi_1, \cdots, \xi_N)$ where a_o and a_1 are positive constants.

AN INEQUALITY FOR THE DIRICHLET PROBLEM

Theorem 4.1. Let u be a function with piecewise continuous second derivatives in B, but otherwise arbitrary, then

$$\int_B u^2 dx \leq \alpha_1 \int_B (\Delta u)^2 dx + \alpha_2 \oint_{\partial B} u^2 dS$$

where α_1 and α_2 are explicitly determined constants [4].

Proof: Introduce an auxiliary function w satisfying

$$\Delta w = u \quad \text{in} \quad B$$

and

$$w = 0 \quad \text{on} \quad \partial B .$$

Then by Green's identity

$$\int_B u^2 dx = \int_B u \, \Delta w \, dx = \oint_{\partial B} u \, \partial w/\partial n \, dS + \int_B w \, \Delta u \, dx \tag{4.1}$$

and Schwarz's inequality implies that

$$\int_B u^2 dx \le \{ \oint_{\partial B} u^2 dS \oint_{\partial B} (\partial w/\partial n)^2 dS\}^{\frac{1}{2}} + \{ \int_B w^2 dx \int_B (\Delta u)^2 dx\}^{\frac{1}{2}} \tag{4.2}$$

We now bound the integrals involving w on the right-hand side of (4.2) in terms of $\int_B u^2 dx$.

We start with

$$\int_B w^2 dx \le \lambda_1^{-1} \int_B w,_i w,_i dx = - \lambda_1^{-1} \int_B w \, \Delta w \, dx \tag{4.3}$$

where we have used (3.2). Using the Schwarz inequality, dividing both sides by $(\int_B w^2 dx)^{\frac{1}{2}}$ and squaring we obtain the first bound:

$$\int_B w^2 dx \le \lambda_1^{-2} \int_B (\Delta w)^2 dx \; . \tag{4.4}$$

Another inequality which we will need later follows from (4.3) and (4.4):

$$\int_B w,_i w,_i dx \le \lambda_1^{-1} \int_B (\Delta w)^2 dx \; . \tag{4.5}$$

To obtain a bound on $\oint_{\partial B} (\partial w/\partial n)^2 dS$ we start with the identity (2.6) of Chapter 2:

$$\begin{aligned} &\oint_{\partial B} [f^i n_i \, \{(\partial u/\partial s)^2 - (\partial u/\partial n)^2\} - 2f^i s_i \, \partial u/\partial n \cdot \partial u/\partial s] \, dS \\ &\quad = - 2 \int_B f^i u,_i \Delta u \, dx + \int_B \{f^i,_i u,_j u,_j - 2f^i,_j u,_i u,_j\} \, dx \; . \end{aligned} \tag{4.6}$$

Putting w into (4.6) we obtain

$$\oint_{\partial B} f^i n_i (\partial w/\partial n)^2 dS = 2 \int_B f^i w,_i \Delta w \, dx - \int_B \{f^k,_k w,_j w,_j -2f^i,_j w,_i w,_j\} dx \tag{4.7}$$

where we have used the fact that $\partial w/\partial s = 0$ on ∂B since $w = 0$ there.

We choose the vector field f^k in such a way that $f^k n_k$ is bounded and has a positive minimum p_m on B. For example, if B is star-shaped with respect to the origin, one can take $f^k = x_k$. A detailed discussion of methods of constructing appropriate vector fields for more general regions is given in [4].

We also have that there exists a constant C such that

$$-\{f^k_{,k}\delta^{ij} - 2f^i_{,j}\}\, w_{,i}w_{,j} \le C\, w_{,i}w_{,i}$$

throughout B if the f^i have bounded first derivatives (C is any upper bound for the largest eigenvalue of the matrix $\{-f^k_{,k}\delta^{ij} + f^i_{,j} + f^j_{,i}\}$). Thus we obtain, from (4.7), that

$$p_m \oint_{\partial B} (\partial w/\partial n)^2 dS \le \alpha \int_B (\Delta w)^2 dx + \left(\frac{|f^i f^i|_M}{\alpha} + C \right) \int_B w_{,i}w_{,i} dx\,, \qquad \alpha > 0 \tag{4.8}$$

where we have used the a-g inequality and $|\cdot|_M$ denotes the maximum of the absolute value of the enclosed quantity. We now obtain the desired bound on $\oint_{\partial B} (\partial w/\partial n)^2 dS$ by combining (4.5) with (4.8):

$$\begin{aligned} \oint_{\partial B} (\partial w/\partial n)^2 dS &\le p_m^{-1} \left\{ \alpha + \lambda_1^{-1} \left(\frac{|f^i f^i|_M}{\alpha} + C \right) \right\} \int_B (\Delta w)^2 dx \\ &\equiv K \int_B u^2 dx. \end{aligned} \tag{4.9}$$

This inequality along with (4.4) and (4.2) then gives

$$\int_B u^2 dx \le \left(K \oint_{\partial B} u^2 dS \int_B u^2 dx\right)^{\frac{1}{2}} + \left(\lambda_1^{-1} \int_B u^2 dx \int_B (\Delta u)^2 dx\right)^{\frac{1}{2}} \tag{4.10}$$

and dividing through by $\int_B u^2 dx$, squaring both sides and using the a-g inequality gives

$$\int_B u^2 dx \le 2\,\lambda_1^{-2} \int_B (\Delta u)^2 \, dx + 2K \oint_{\partial B} u^2 dS \; .$$

Thus the coefficients α_1, α_2 in the statement of the theorem are

$$\alpha_1 = 2\,\lambda_1^{-2}$$

$$\alpha_2 = 2\,p_m^{-1} \, \{\alpha + \lambda_1^{-1} (\alpha^{-1} \, |f^i f^i|_M + C)\}, \qquad \alpha > 0 \; .$$

AN ALTERNATIVE APPROACH

As mentioned earlier, we are carrying out the derivations for some of the more simple cases but in a manner which immediately extends to more general operators. If we had not been concerned with techniques which apply to more general elliptic operators we could have derived an inequality like that of Theorem 4.1 more simply as follows:

Decompose u as

$$u = h + g$$

where $\Delta h = 0$ on B, $g = 0$ on ∂B so that $u = h$ on ∂B, $\Delta u = \Delta g$ in B. Then squaring both sides, integrating and using the a-g inequality we have

$$\int_B u^2 dx \le 2 \int_B h^2 dx + 2 \int_B g^2 dx \; .$$

Then from (3.1) and (3.6) we have

$$\int_B u^2 dx \le 2\, q_1^{-1} \oint_{\partial B} u^2 dS + 2\,\lambda_1^{-2} \int_B (\Delta u)^2 dx \; .$$

Theorem 4.2. Let ψ be piecewise $C^2(\bar{B})$ and define $u = \psi + c$ where $c = -\frac{1}{\omega_N a^{N-1}} \oint_{\partial S_a} \psi \, dS$, $S_a \subset B$ denotes the interior of a sphere of radius a centered at the origin (we assume $0 \in S$) and ω_N is the surface area of S_a. Then

$$\int_B u^2 dx \le \alpha_1 \int_B (\Delta u)^2 dx + \alpha_2 \oint_{\partial B} (\partial u/\partial n)^2 dS$$

where α_1 and α_2 are coefficients which will be explicitly determined [3].

Proof: The derivation of this inequality is an interesting variation of the preceding. It is slightly complicated by the compatibility condition which solutions of the Neumann problem must satisfy. Notice that by definition $\oint_{\partial B} u \, dS = 0$.

We start with Green's identity

$$\int_B u,_i u,_i dx = \oint_{\partial B} u \, \partial u/\partial n \, dS - \int_B u \, \Delta u \, dx$$

and apply the Schwarz inequality

$$\int_B u,_i u,_i dx \le \left(\oint_{\partial B} pu^2 dS \right)^{\frac{1}{2}} \left(\oint_{\partial B} p^{-1} (\partial u/\partial n)^2 dS \right)^{\frac{1}{2}} + \left(\int_B u^2 dx \right)^{\frac{1}{2}} \left(\int_B (\Delta u)^2 dx \right)^{\frac{1}{2}}, \qquad p > 0 . \tag{4.11}$$

We now bound $\oint_{\partial B} pu^2 dS$ and $\int_B u^2 dx$ in terms of $\int_B u,_i u,_i dx \equiv D(u,u)$.

Denote by B_a the region $B - \bar{S}_a$ and let f^i be a sufficiently smooth vector field defined in $\bar{B}_a$. Then by the divergence theorem we have

$$\oint_{\partial B} f^i n_i u^2 dS = - \oint_{\partial S_a} f^i n_i u^2 dS + \int_{B_a} f^i,_i u^2 dx + 2 \int_{B_a} f^i u \, u,_i dx . \tag{4.12}$$

An application of the a-g mean inequality applied to the last term on the right of (4.12) yields

$$\oint_{\partial B} f^i n_i u^2 dS \le - \oint_{\partial S_a} f^i n_i u^2 dS + \int_{B_a} (f^i_{,i} + \frac{1}{\alpha} f^i f^i)\, u^2 dx + \int_{B_a} \alpha\, u_{,i} u_{,i} dx \tag{4.13}$$

where α is some positive function in B_a.

We assume now that f^i and α have been chosen so that

$$\begin{aligned} p \equiv f^i n_i &\ge K_1 > 0 \quad \text{on} \quad \partial B\,, \\ - f_i n_i &\le K_2 \quad \text{on} \quad \partial S_a, \\ f^i_{,i} + \frac{1}{\alpha} f^i f^i &\le 0 \quad \text{in} \quad B_a\,, \end{aligned} \tag{4.14}$$

where K_1 and K_2 are constants (see [3] for details).

Using (4.14) with (4.13) we have that

$$\oint_{\partial B} u^2 dS \le K_2 \oint_{\partial S_a} u^2 dS + \bar{\alpha} \int_{B_a} u_{,i} u_{,i} dx \tag{4.15}$$

where $\bar{\alpha}$ is an upper bound for α in B_a. Now since $\oint_{\partial S_a} u\, dS = 0$ then by (3.8)

$$\oint_{\partial S_a} u^2 dS \le p_2^{-1} \int_{S_a} u_{,i} u_{,i} dx = a \int_{S_a} u_{,i} u_{,i} dx \tag{4.16}$$

since for the sphere of radius a, $p_2 = 1/a$, (see (3.27)).

Combining (4.15) and (4.16) it follows that

$$\oint_{\partial B} p u^2 dS \le K_3 \cdot D(u,u) \tag{4.17}$$

where $K_3 = \max\,(a \cdot K_2, \bar{\alpha})$, or using (4.14),

$$\oint_{\partial B} u^2 dS \le (K_3 / K_1)\, D(u,u)\,. \tag{4.18}$$

Now from the divergence theorem

$$\oint_{\partial B} x_i n_i u^2 \, dS = N \int_B u^2 \, dx + 2 \int_B x_i u \, u_{,i} \, dx \, . \tag{4.19}$$

Using the a-g mean inequality it follows easily that

$$\int_B u^2 \, dx \le \frac{2r_M}{N} \oint_{\partial B} u^2 \, dS + \frac{4r_M}{N^2} D(u,u) \tag{4.20}$$

where r_M is the maximum distance from the origin to ∂B. This inequality with (4.18) yields

$$\int_B u^2 \, dx \le K_4 \cdot D(u,u) \, , \tag{4.21}$$

where

$$K_4 = \frac{2r_M}{N} \left[K_3 / K_1 + \frac{2r_M}{N} \right] .$$

We now obtain the desired inequality by using (4.17) and (4.21) in (4.11) to obtain

$$D(u,u)^{\frac{1}{2}} \le K_3^{\frac{1}{2}} \left(\oint_{\partial B} p^{-1} (\partial u / \partial n)^2 \, dS \right)^{\frac{1}{2}} + K_4^{\frac{1}{2}} \left(\int_B (\Delta u)^2 \, dx \right)^{\frac{1}{2}}$$

or, squaring and using the a-g mean inequality

$$D(u,u) \le 2K_3 \oint_{\partial B} p^{-1} (\partial u / \partial n)^2 \, dS + 2K_4 \int_B (\Delta u)^2 \, dx \, .$$

Finally, by (4.21)

$$\int_B u^2 \, dx \le (2K_3 K_4 / K_1) \oint_{\partial B} (\partial u / \partial n)^2 \, dS + 2K_4^2 \int_B (\Delta u)^2 \, dx \, .$$

THE ROBIN PROBLEM

Theorem 4.3. Let u be a function with piecewise continuous second derivatives in B, but otherwise arbitrary, then

$$\int_B u^2 dx \le \alpha_1 \int_B (\Delta u)^2 dx + \alpha_2 \oint_{\partial B} (\partial u/\partial n + \alpha u)^2 dS$$

where $\alpha = \alpha(x)$ is a positive, piecewise continuous function which is bounded away from zero on ∂B and α_1 and α_2 are explicitly determined constants [7].

Proof: Let a function w be defined such that

$$\Delta w = u \quad \text{in} \quad B$$

$$\frac{\partial w}{\partial n} + \alpha w = 0 \quad \text{on} \quad \partial B .$$

Then we have

$$\int_B u^2 dx = \int_B u\, \Delta w\, dx = - \oint_{\partial B} w(\partial u/\partial n + \alpha u) dS + \int_B w\, \Delta u\, dx . \tag{4.22}$$

We now decompose the function u into the sum of two functions h and g which satisfy

$$\Delta h = 0 \quad \text{in} \quad B , \qquad\qquad \Delta g = \Delta u \quad \text{in} \quad B ,$$

$$\partial h/\partial n + \alpha h = \partial u/\partial n + \alpha u \quad \text{on} \quad \partial B , \quad \partial g/\partial n + \alpha g = 0 \quad \text{on} \quad \partial B .$$

Now since (4.22) holds for any sufficiently regular function u it holds for $u \equiv h$ and from (4.22) it follows that

$$\int_B h^2 dx = - \oint_{\partial B} w(\partial h/\partial n + \alpha h) dS$$

and the Schwarz inequality yields

$$\left(\int_B h^2 dx\right)^2 \le \oint_{\partial B} \alpha\, w^2 dS \oint_{\partial B} \alpha^{-1} (\partial h/\partial n + \alpha h)^2 dS . \tag{4.23}$$

If we denote by K a lower bound for the first eigenvalue in the elastically supported membrane problem [29] we obtain

$$\int_B w^2 dx \le K^{-1} \left[D(w,w) + \oint_{\partial B} \alpha\, w^2 dS\right] \le \int_B (\Delta w)^2 dx \tag{4.24}$$

since

$$K \le \frac{D(w,w) + \alpha_m \oint_{\partial B} w^2 dS}{\int_B w^2 dx} \le \frac{D(w,w) + \oint_{\partial B} \alpha\, w^2 dS}{\int_B w^2 dx}$$

where α_m is the greatest lower bound for α on ∂B.

Now from (4.24) it follows that

$$\oint_{\partial B} \alpha\, w^2 dS \le K^{-1} \int_B (\Delta w)^2 dx \ . \tag{4.25}$$

Inserting (4.25) into (4.23) we have

$$\int_B h^2 dx \le K^{-1} \oint_{\partial B} \alpha^{-1} (\partial h/\partial n + \alpha h)^2 dS \ . \tag{4.26}$$

Now take $u \equiv g$. Then from (4.22) it follows that

$$\int_B (\Delta w)^2 dx = \int_B g^2 dx = \int_B w\, \Delta g\; dx \ .$$

From Schwarz's inequality we have

$$\int_B (\Delta w)^2 dx \le \int_B w^2 dx \int_B (\Delta g)^2 dx$$

and from (4.24) we have

$$\int_B g^2 dx \le K^{-2} \int_B (\Delta g)^2 dx \ . \tag{4.27}$$

Hence we have from (4.26) and (4.27) using the representation $u = h + g$

$$\int_B u^2 dx \le 2 \left\{K^{-1} \oint_{\partial B} \alpha^{-1} (\partial u/\partial n + u\alpha)^2 dS + K^{-2} \int_B (\Delta u)^2 dx\right\}$$

which is valid for any sufficiently smooth function u.

AN INEQUALITY FOR A PROBLEM IN ELASTICITY

In this section we derive an a priori inequality which has applications to the first boundary value problem in elasticity

$$L_i(u) \equiv u_{i,jj} + \alpha_{j,ji} = F_i \,, \quad \text{in} \quad B$$
$$u = f_i \,, \quad \text{on} \quad \partial B \qquad i=1,2,\cdots,N \,. \tag{4.28}$$

In this context u is the displacement vector with components u_i which satisfy the system (4.28), α involves the elastic constants λ and μ ($\alpha = (\lambda + \ )/\mu = (1-2\sigma)^{-1}$, σ denoting Poisson's ratio), and the F_i are proportional to the body force. The a priori inequality of interest is then given by the following

Theorem 4.4. Let $u = (u_1, u_2, \cdots, u_N)$ be an arbitrary vector field with piecewise continuous second derivatives in B. Then

$$\int_B u_i u_i dx \le \alpha_1 \int_B L_i(u)\, L_i(u) dx + \alpha_2 \oint_{\partial B} \{u_i u_i + (u_i n_i)^2\}\, dS$$

where α_1 and α_2 are explicitly determined constants [2].

Proof: We must first derive an auxiliary inequality. Let ψ_i be the ith component of a vector function which possesses piecewise continuous second derivatives in B and vanishes on ∂B. Then, as in the derivation of (2.6), an application of Green's first identity and integration by parts results in the identity

$$2\int_B f^k \psi_{i,k} L_i(\psi) dx = 2 \oint_{\partial B} f^k \psi_{i,k}[\psi_{i,j} n_j + \alpha\psi_{j,j} n_i] dS$$
$$+ \int_B f^k_{,k}[\psi_{i,j}\psi_{i,j} + \alpha\psi_{i,i}\psi_{j,j}]dx - 2\int_B [f^k_{,j}\psi_{i,k}\psi_{i,j} + \alpha f^k_{,i}\psi_{i,k}\psi_{j,j}]dx$$
$$- \oint_{\partial B} f^k n_k [\psi_{i,j}\psi_{i,j} + \alpha\psi_{i,i}\psi_{j,j}]dS \,.$$

Since ψ_i vanishes on ∂B the first and last integrals on the right combine to give

$$\oint_{\partial B} f^k n_k[\psi_{i,j}\psi_{i,j} + \alpha\psi_{i,i}\psi_{j,j}]dS = 2\int_B f^k\psi_{i,k}L_i(\psi)dx$$

$$-\int_B f^k_{,k}[\psi_{i,j}\psi_{i,j}+\alpha\psi_{i,i}\psi_{j,j}]dx + 2\int_B [f^k_{,j}\psi_{j,k}\psi_{i,j}+\alpha f^k_{,i}\psi_{i,k}\psi_{j,j}]dx \ .$$

Since f^i has bounded first derivatives we may easily obtain, for α positive

$$2\int_B [f^k_{,j}\psi_{i,k}\psi_{i,j} + \alpha f^k_{,i}\psi_{i,k}\psi_{j,j}]dx - \int_B f^k_{,k}[\psi_{i,j}\psi_{i,j} + \alpha\psi_{i,i}\psi_{j,j}]dx$$

$$\leq b\int_B [\psi_{i,j}\psi_{i,j} + \alpha\psi_{i,i}\psi_{j,j}]dx$$

where the constant b may be easily obtained. By the a-g inequality and the fact that α is positive, we have

$$\oint_{\partial B} p[\psi_{i,j}\psi_{i,j}+\alpha\psi_{i,i}\psi_{j,j}]dS \leq (b+|f^k f^k|_M \beta^{-1})\int_B [\psi_{i,j}\psi_{i,j}+\alpha\psi_{i,i}\psi_{j,j}]dx$$

$$+ \beta\int_B L_i(\psi)\ L_i\psi\ dx \qquad (4.29)$$

for any positive β. We have again used the definition $p = f^k n_k$ and chosen the f^k such that $p > 0$ on ∂B. Now since ψ_i vanishes on ∂B we have

$$\int_B [\psi_{i,j}\psi_{i,j} + \alpha\psi_{i,i}\psi_{j,j}]dx = -\int_B \psi_i L_i\psi\ dx \ ,$$

or, by Schwarz's inequality

$$\frac{\int_B [\psi_{i,j}\psi_{i,j} + \alpha\psi_{i,i}\psi_{j,j}]dx}{\int_B \psi_i\psi_i dx} \leq \frac{\int_B L_i(\psi)\ L_i(\psi)\ dx}{\int_B (\psi_{i,j}\psi_{i,j} + \alpha\psi_{i,i}\psi_{j,j})dx} \ .$$

Moreover, since $\alpha > 0$

$$\frac{\int_B [\psi_{i,j}\psi_{i,j} + \alpha\psi_{i,i}\psi_{j,j}]dx}{\int_B \psi_i\psi_i dx} \geq \frac{\int_B \psi_{i,j}\psi_{i,j} dx}{\int_B \psi_i\psi_i dx} \geq \lambda_1$$

by (3.2). Thus the preceding two inequalities give us that

$$\int_B \psi_i\psi_i dx \leq \lambda_1^{-1} \int_B [\psi_{i,j}\psi_{i,j} + \alpha\psi_{i,i}\psi_{j,j}]dx \leq \lambda_1^{-2} \int_B L_i\psi\ L_i\psi\ dx . \qquad (4.30)$$

Use of this inequality in (4.29) then yields

$$\oint_{\partial B} p[\psi_{i,j}\psi_{i,j} + \alpha\psi_{i,i}\psi_{j,j}]dS \leq [b\lambda_1^{-1} + 2\ \{|f^k f^k|_M\ \lambda_1^{-1}\}^{\frac{1}{2}} \int_B L_i\psi\ L_i\psi\ dx \qquad (4.31)$$

where we have made the optimal choice for β.

We are now ready to determine a bound for $\int_B u_i u_i dx$. To this end we introduce the auxiliary vector χ with components χ_i defined by

$$\begin{aligned} L_i(\chi) &= u_i \quad \text{in} \quad B \\ & \qquad\qquad\qquad i=1,2,\cdots,N \\ \chi_i &= 0 \quad \text{on} \quad \partial B . \end{aligned}$$

Then

$$\int_B u_i u_i dx = \int_B u_i L_i(\chi) dx = \oint_{\partial B} u_i[\chi_{i,j}n_j + \alpha\chi_{j,j}n_i]dS + \int_B \chi_i L_i u\ dx .$$

By Schwarz's inequality for vectors we have

$$\int_B u_i u_i dx = \int_B L_i(\chi)L_i(\chi)dx \leq \{\oint_{\partial B} p^{-1}[u_i u_i + (u_i n_i)^2]dS \oint_{\partial B} p[\chi_{i,j}\chi_{i,j}$$

$$+ \alpha\chi_{i,i}\chi_{j,j}dS]^{\frac{1}{2}} + \{\int_B \chi_i\chi_i dx \int_B L_i(u)\ L_i(u)dx\}^{\frac{1}{2}} .$$

Since $\chi_i = 0$ on ∂B we may use (4.30) and (4.31) and the a-g mean inequality with ψ_i replaced by χ_i to obtain

$$\int_B u_i u_i dx \le 2p_m^{-1}[b\lambda_1^{-1} + 2(|f^k f^k|_M \lambda_1^{-1})^{\frac{1}{2}}] \oint_{\partial B} [u_i u_i + (u_i n_i)^2] dS$$

$$+ 2\lambda_1^{-2} \int_B L_i(u)\, L_i(u)\, dx .$$

5 *A priori* inequalities II – Second order parabolic applications

The *a priori* inequalities presented in this chapter have applications to the Dirichlet, Neumann and Robin problems for second order parabolic equations. Although they are derived in terms of the heat operator $Lu = \Delta u - u_t$, the derivations can be carried over to the general second order operator $\mathcal{L}u = (a_{ij}u_{,i})_{,j} - c(x)u_t$, where $c(x) > 0$, $a_{ij} = a_{ij}(x,t)$ is a symmetric matrix such that $a_0\xi_i\xi_i \le a_{ij}\xi_i\xi_i \le a_1\xi_i\xi_i$ for all $(x,t) \in D$ and all real vectors $\xi = (\xi_1, \cdots, \xi_N)$ and a_0 and a_1 are positive constants. Inequalities applicable to semi-linear problems are also given.

The approach used in the derivation of the inequality for the parabolic Dirichlet problem has some similarly to that of the corresponding elliptic case but for the other problems new techniques are needed.

AN INEQUALITY FOR THE DIRICHLET PROBLEM

Theorem 5.1. Let u be a function with piecewise continuous second derivatives in x, and piecewise continuous first derivatives in t throughout D, but otherwise arbitrary. Then

$$\int_D u^2\,dV \le \alpha_1 \int_D (Lu)^2\,dV + \alpha_2 \int_B u^2\,dx + \alpha_3 \int_S u^2\,d\sigma$$

where α_1, α_2 and α_3 are explicitly determined constants [39].

Proof: Introduce the auxiliary function w which satisfies the boundary value problem

$$L^*w \equiv \Delta w + w_t = u \quad \text{in} \quad D \cup B$$

$$w = 0 \qquad\qquad \text{on} \quad B_T \cup \bar{S}\ .$$

Using the divergence theorem and Green's second identity we can write

$$\int_D u^2 dV = \int_D u\, L^* w\ dV = \int_S u\ \partial w/\partial n\ d\sigma + \int_D w\ Lu\ dV - \int_B u\ w\ dx\ .$$

The vector form of the Schwarz inequality then yields

$$\int_D u^2 dV \le \left(\int_S u^2 d\sigma \int_S (\partial w/\partial n)^2 d\sigma\right)^{\frac{1}{2}} + \left(\int_B w^2 dx + \lambda_1 \int_D w^2 dV\right)^{\frac{1}{2}} \cdot \left(\int_B u^2 dx + \lambda_1^{-1} \int_D (Lu)^2 dV\right)^{\frac{1}{2}} \tag{5.1}$$

Our object now is to bound

$$\int_B w^2 dx + \lambda_1 \int_D w^2 dV \tag{5.2}$$

and

$$\int_S (\partial w/\partial n)^2 d\sigma \tag{5.3}$$

in terms of $\int_D u^2 dV$.

To obtain the bound on (5.2) we use the fact that

$$-\int_B w^2 dx = \int_D \partial w^2/\partial t\ dV = 2\int_D w[L^* w - w\ \Delta\ w]\ dV$$

and recalling that $w = 0$ on S we obtain

$$\int_B w^2 dx + 2\int_D w,_i w,_i dV = -\int_D w\ L^*\ w\ dV\ . \tag{5.4}$$

This equation will be useful very shortly; it also yields (using (3.2) integrated over [0,T]) the important inequality

$$\int_D w,_i w,_i dV \le \lambda_1^{-1} \int_D (L^* w)^2 dV\ . \tag{5.5}$$

The desired bound on (5.2) is now obtained as follows:

$$\int_B w^2 dx + \lambda_1 \int_D w^2 dV \le \int_B w^2 dx + 2\int_D w,_i w,_i dV - \lambda_1 \int_D w^2 dV$$
$$\le \lambda_1^{-1} \int_D (L^* w)^2 dV = \lambda_1^{-1} \int_D u^2 dV \tag{5.6}$$

where we have used (5.4) and the a-g inequality.

To obtain the bound on (5.3) we use Equation (2.7) recalling that $\partial w/\partial s = \partial w/\partial t = 0$ on S and $\int_{B_T} f^{N+1} n_{N+1} w,_i w,_i dx = 0$ since $w = 0$ on $S \cup B_T$:

$$\int_S f^i n_i (\partial w/\partial n)^2 d\sigma = 2\int_D f^\alpha w,_\alpha L^* w \, dV - \int_D \{f^\alpha,_\alpha w,_i w,_i - 2f^j,_i w,_i w,_j\} dV$$
$$+ 2\int_D f^{N+1},_i w_t w,_i dV - 2\int_D f^\alpha w,_\alpha w_t dV - \int_B f^{N+1} w,_i w,_i dx \tag{5.7}$$

Now choose $f^{N+1} \ge 0$ in $D \cup B$, then the last term in (5.7) is nonpositive and we drop it. Further, the term next to it contains the term $-2\int_D f^{N+1}(w_t)^2 dV \le 0$ and we use the a-g and Schwarz inequalities on the other integrals containing w_t to cancel out this term. We thus obtain the inequality

$$\int_S f^i n_i (\partial w/\partial n)^2 d\sigma \le \bar{K} \int_D (L^* w)^2 dV = \bar{K} \int_D u^2 dV$$

where

$$\bar{K} = 1 + |f^{N+1}|_M + \lambda_1^{-1} \left\{ |f^i f^i|_M^{\frac{1}{2}} + 2 \left| \frac{f^{N+1},_i f^{N+1},_i}{f^{N+1}} \right|_M + 2 \left| \frac{f^i f^i}{f^{N+1}} \right|_M + C \right\},$$

and where C is an upper bound on the largest eigenvalue of the matrix $-\{f^\alpha,_\alpha \delta_{ij} - f^j,_i - f^i,_j\}$. If we now chose the f^ℓ, $\ell = 1, \cdots, N$ such that $f^\ell n_\ell$ has a positive minimum p_m on S we have

$$\int_S (\partial w/\partial n)^2 d\sigma \le K \int_D u^2 dV \tag{5.8}$$

where $K = p_m^{-1} \bar{K}$.

The inequalities (5.6) and (5.8) thus yield the inequality of the theorem where

$$\alpha_1 = 2\lambda_1^{-2}, \qquad \alpha_2 = 2\lambda_1^{-1}, \qquad \alpha_3 = 2K .$$

A SEMI-LINEAR CASE

We now establish an inequality for a function which satisfies the semi-linear equation $Lu = f(x,t,u)$.

Theorem 5.2. Let u and φ be functions defined in D, piecewise C^2 in x and piecewise C^1 in t. Furthermore we assume that u satisfies the semi-linear equation $Lu = f(x,t,u)$ and that f satisfies a Lipschitz condition in u with Lipschitz constant M. Let $\psi(x,t) = u(x,t) - \varphi(x,t)$. Then

$$\int_D \psi^2 dV \le \alpha_1 \int_B \psi^2 dx + \alpha_2 \int_S \psi^2 d\sigma + \alpha_3 \int_D F^2 dV$$

where

$$F(x,t) = f(x,t,\varphi) - L\varphi$$

and α_1, α_2, α_3 are explicitly determined constants [39].

Proof: Introduce the function

$$v = \psi\, e^{b(T-t)}$$

in D where b is a positive constant. Then

$$L\psi = (Lv - bv)\, e^{-b(T-t)} \tag{5.9}$$

and we seek a bound on

$$\int_D e^{2b(T-t)} \psi^2 dV = \int_D v^2 dV \ . \tag{5.10}$$

To this end introduce the function w which satisfies the boundary value problem

$$L^*w - bw = v \quad \text{in} \quad D \cup B$$

$$w = 0 \qquad\qquad \text{on} \quad \bar{B}_T \cup S \ .$$

Now much of the remainder of the derivation proceeds in a manner parallel to that of the preceding theorem and so we will just give the highlights. Proceeding then as in Theorem 5.1, we obtain the inequality

$$\begin{aligned} \int_D v^2 dV \le \left(\int_S v^2 d\sigma \int_S (\partial w/\partial n)^2 d\sigma\right)^{\frac{1}{2}} + \left(\int_B w^2 dx + (b+\lambda_1) \int_D w^2 dV\right)^{\frac{1}{2}} \\ \cdot \left(\int_B v^2 dx + (b+\lambda_1)^{-1} \int_D (Lv-bv)^2 dV\right)^{\frac{1}{2}} \end{aligned} \tag{5.11}$$

and thus we need bounds on $\int_B w^2 dx + (b+\lambda_1) \int_D w^2 dV$ and $\int_S (\partial w/\partial n)^2 d\sigma$.

The first bound is obtained from the easily derived identity

$$\int_B w^2 dx + 2 \int_D w,_i w,_i dV + 2b \int_D w^2 dV = - 2 \int_D w(L^*w-bw) \ dV \ . \tag{5.12}$$

Using (5.12) and (3.2) along with the a-g inequality we obtain

$$\int_B w^2 dx + (b+\lambda_1) \int_D w^2 dV \le (b+\lambda_1)^{-1} \int_D v^2 dV \ . \tag{5.13}$$

Further, (5.12) also yields the useful inequality

$$\int_D w,_i w,_i dV \le (4b)^{-1} \int_D v^2 dV \ . \tag{5.14}$$

To obtain the bound on $\int_S (\partial w/\partial n)^2 d\sigma$ we modify (5.7) in an obvious way

$$\int_S f^i n_i (\partial w/\partial n)^2 d\sigma = 2\int_D f^\alpha w_{,\alpha}(L^* w - bw)\,dV - \int_D \{f^\alpha_{,\alpha} w_{,i} w_{,i} - 2f^j_{,i} w_{,i} w_{,j}\}\,dV$$

$$+ 2\int_D f^{N+1}_{,i}\, w_t w_{,i}\,dV - 2\int_D f^\alpha w_{,\alpha} w_t\,dV - \int_B f^{N+1} w_{,i} w_{,i}\,dx$$

$$- b\int_D f^\alpha_{,\alpha} w^2\,dV - b\int_B f^{N+1}\, w^2\,dx \quad .$$

Now choosing $f^{N+1} \geq 0$ in $\bar{D}$ and the f^ℓ such that $f^\ell n_\ell$ has a positive minimum p_m on S we have

$$p_m \int_S (\partial w/\partial n)^2 d\sigma \leq 2\int_D f^\alpha w_{,\alpha}(L^* w - bw)\,dV - \int_D \{f^\alpha_{,\alpha} w_{,i} w_{,i} - 2f^j_{,i} w_{,i} w_{,j}\}\,dV$$

$$+ 2\int_D f^{N+1}_{,i}\, w_t w_{,i}\,dV - 2\int_D f^\alpha w_{,\alpha} w_t\,dV - b\int_D f^\alpha_{,\alpha} w^2\,dV$$

from which follows the inequality

$$\int_S (\partial w/\partial n)^2 d\sigma \leq K \int_D v^2\,dV \tag{5.15}$$

where

$$K = p_m^{-1} \left\{ 1 + |f^{N+1}|_M + b\,|f^\alpha_{,\alpha}|_M (b+\lambda_1)^{-2} + (4b)^{-1}\left[|f^\ell f^\ell|_M^{\frac{1}{2}} + 2\left|\frac{f^{N+1}_{,i}\; f^{N+1}_{,i}}{f^{N+1}}\right|_M + 2\left|\frac{f^i f^i}{f^{N+1}}\right|_M + C \right] \right\} \quad , \tag{5.16}$$

where again C is any upper bound on the largest eigenvalue of the matrix $-\{f^\alpha_{,\alpha}\,\delta_{ij} - f^j_{,i} - f^i_{,j}\}$.

From the Lipschitz condition satisfied by f we have

$$|L\psi| \leq M\,|\psi| + |F| \tag{5.17}$$

so that by (5.9) we have

$$\int_D (Lv-bv)^2 dV = \int_D e^{2b(T-t)} (L\psi)^2 dV$$

$$\leq 2M^2 \int_D e^{2b(T-t)} \psi^2 dV + 2 \int_S e^{2b(T-t)} F^2 dV$$

and this inequality along with (5.13) and (5.15) in (5.11) gives

$$\left(\int_D v^2 dV\right)^{\frac{1}{2}} \leq \left(K \int_S v^2 d\sigma\right)^{\frac{1}{2}}$$

$$+ \left[(b+\lambda_1)^{-1} \left\{\int_B v^2 dx + (b+\lambda_1)^{-1}\left(2M^2 \int_D v^2 dV + 2 \int_D e^{2b(T-t)} F^2 dV\right)\right\}\right]^{\frac{1}{2}}$$

or

$$\left(\int_D v^2 dV\right)^{\frac{1}{2}} \leq \left(K \int_S v^2 d\sigma\right)^{\frac{1}{2}} + \left[(b+\lambda_1)^{-1} \int_B v^2 dx + (b+\lambda_1)^{-2} \int_D e^{2b(T-t)} F^2 dV\right]^{\frac{1}{2}}$$

$$+ \sqrt{2}\, M\, (b+\lambda_1)^{-1} \left(\int_D v^2 dV\right)^{\frac{1}{2}}$$

where we have used the inequality $(c+d)^{\frac{1}{2}} \leq \sqrt{c} + \sqrt{d}$, $c, d \geq 0$. Now choosing b such that $\tilde{K} \equiv 1 - M\sqrt{2}\,(b+\lambda_1)^{-1} > 0$ we obtain the inequality

$$\int_D v^2 dV \leq 2\, \tilde{K}^{-2} \left\{K \int_S v^2 d\sigma + (b+\lambda_1)^{-1} \int_B v^2 dx + (b+\lambda_1)^{-2} \int_D e^{2b(T-t)} F^2 dV\right\}.$$

AN INEQUALITY FOR THE NEUMANN PROBLEM

Theorem 5.3. Let u be a function with piecewise C^2 in x, and piecewise C^1 in t, but otherwise arbitrary. Then

$$\int_D u^2 dV \leq \alpha_1 \int_D (Lu)^2 dV + \alpha_2 \int_B u^2 dx + \alpha_3 \int_S (\partial u/\partial n)^2 d\sigma$$

where α_1, α_2 and α_3 are explicitly determined constants [19].

Proof: Decompose u as $u = f + g + h$ where

$$
\begin{array}{llll}
Lf = 0, & Lg = Lu, & Lh = 0 & \text{in } D, \\
f = u, & g = 0, & h = 0 & \text{on } B, \\
\partial f/\partial n = 0, & \partial g/\partial n = 0, & \partial h/\partial n = \partial u/\partial n & \text{on } S.
\end{array}
$$

and successively substitute f, g and h into the identity

$$
\begin{aligned}
\int_{B_t} w^2 dx = \int_B w^2 dx - 2\int_0^t \int_{B_\tau} w\, L\, w\, dx\, d\tau - 2\int^t \int_{B_\tau} w_{,i} w_{,i} dx\, d\tau \\
+ 2\int^t \int_{\partial B_\tau} w\, \partial w/\partial n\, ds\, d\tau \quad .
\end{aligned}
\tag{5.18}
$$

Putting f into (5.18) yields

$$\int_{B_t} f^2 dx \le \int_B u^2 dx \ ,$$

and integrating from $t = 0$ to $t = T$ gives

$$\int_D f^2 dx\, dt \le T\int_B u^2 dx \ . \tag{5.19}$$

Putting g into (5.18) yields

$$
\begin{aligned}
\int_{B_t} g^2 dx &\le -2\int_0^t \int_{B_\tau} g\, Lu\, dx\, d\tau \\
&\le \alpha \int_0^t \int_{B_\tau} g^2 dx\, d\tau + \alpha^{-1} \int_0^t \int_{B_\tau} (Lu)^2\, dx\, d\tau
\end{aligned}
$$

by the arithmetic-geometric mean inequality for arbitrary positive α. Multiplying by $e^{-\alpha t}$ and rearranging gives

$$\frac{d}{dt}\left(e^{-\alpha t}\int_0^t \int_{B_t} g^2 dx\, d\tau\right) \le \alpha^{-1} e^{-\alpha t}\int_0^t \int_{B_\tau} (Lu)^2 dx\, d\tau \quad .$$

Integration with respect to t from 0 to T and multiplication by $e^{\alpha T}$ then gives

$$\int_D g^2 dx\ dt \le \alpha^{-1}\ e^{\alpha T} \int_0^T e^{-\alpha t} \int_0^t \int_{B_t} (Lu)^2 dx\ d\tau\ dt$$

$$= \alpha^{-2} \int_0^T (e^{\alpha(T-t)} - 1) \int_{B_t} (Lu)^2 dx\ dt$$

$$\le \alpha^{-2}\ (e^{\alpha T} - 1) \int_D (Lu)^2 dx\ dt\ ,$$

where integration by parts was used. Setting $\alpha = \beta T^{-1}$ gives

$$\begin{aligned}\int_D g^2 dx\ dt &\le \beta^{-2}\ (e^{\beta}-1)\ T^2 \int_D (Lu)^2 dx\ dt \\ &\le 1.544138653\ T^2 \int_D (Lu)^2 dx\ dt\ ,\end{aligned} \tag{5.20}$$

with the optimal choice of $\beta = 1.59362$.

Finally, putting h into (5.18) yields

$$\begin{aligned}\int_{B_\tau} h^2 dx &= -\ 2 \int_0^t \int_{B_\tau} h,_i h,_i dx\ d\tau + 2 \int_0^t \int_{S_\tau} h \frac{\partial u}{\partial n} ds\ d\tau \\ &\le -\ 2 \int_0^t \int_{B_\tau} h,_i h,_i dx\ d\tau + \alpha^{-1} \int_0^t \int_{S_\tau} \left(\frac{\partial u}{\partial n}\right)^2 ds\ d\tau \\ &\quad + \alpha \int_0^t \int_{S_\tau} h^2 ds\ d\tau\ .\end{aligned} \tag{5.21}$$

Suppose B is star shaped with respect to a point which we take to be the origin. Then $x_i n_i > 0$ on the ∂B. Let $d = \min n_i x_i$, $r^2 = \max x_i x_i$. Then

$$\int_{S_\tau} h^2 dx \le d^{-1} \int_{S_\tau} h^2 x_i n_i ds = d^{-1} \int_{B_\tau} (h^2 x_i),_i\ dx$$

$$= N\ d^{-1} \int_{B_\tau} h^2 dx + 2\ d^{-1} \int_{B_\tau} h\ h,_i x_i dx$$

$$\le N\ d^{-1} \int_{B_\tau} h^2 dx + 2\ d^{-1}\ (r^2 \int_{B_\tau} h^2 dx \int_{B_\tau} h,_i h,_i dx)^{\frac{1}{2}}$$

$$\le N\ d^{-1} \int_{B_\tau} h^2 dx + 2\ \alpha^{-1} \int_{B_\tau} h,_i h,_i dx + \frac{1}{2}\ \alpha\ r^2\ d^{-2} \int_{B_\tau} h^2 dx\ .$$

Putting this into (5.21) gives

$$\int_{B_\tau} h^2 dx \le \alpha^{-1} \int_0^t \int_{S_\tau} \left(\frac{\partial u}{\partial n}\right)^2 ds\, d\tau + \alpha(nd^{-1} + \frac{1}{2}\alpha\, r^2 d^{-2}) \int_0^t \int_{B_\tau} h^2 dx\, d\tau \quad (5.22)$$

Define $K(\alpha) = \alpha(nd^{-1}+\frac{1}{2}\alpha\, r^2 d^{-2})$. Multiplying (5.22) by $e^{-K(\alpha)t}$ and rearranging gives

$$\frac{d}{dt}\left(e^{-K(\alpha)t} \int_0^t \int_{B_\tau} h^2 dx\, d\tau\right) \le \alpha^{-1}\, e^{-K(\alpha)t} \int_0^t \int_{S_\tau} \left(\frac{\partial u}{\partial n}\right)^2 ds\, d\tau .$$

Integrating with respect to t from 0 to T and multiplying by $e^{\alpha T}$ yields

$$\int_D h^2 dx\, dt \le \alpha^{-1}\, e^{K(\alpha)T} \int_0^t e^{-K(\alpha)t} \int_0^t \int_{S_\tau} \left(\frac{\partial u}{\partial n}\right)^2 ds\, d\tau\, dt$$

$$= (\alpha K(\alpha))^{-1} \int_0^T (e^{K(\alpha)(T-t)} - 1) \int_{S_t} \left(\frac{\partial u}{\partial n}\right)^2 ds\, dt$$

$$\le (\alpha K(\alpha))^{-1} (e^{K(\alpha)T} - 1) \int_S \left(\frac{\partial u}{\partial n}\right)^2 ds\, dt .$$

For given region and T, minimize $(\alpha K(\alpha))^{-1} \cdot (e^{K(\alpha)T} - 1)$ with respect to α, to obtain

$$\int_D h^2 dx\, dt \le [c_2(R,T)]^2 \int_S \left(\frac{\partial u}{\partial n}\right)^2 ds\, dt . \quad (5.23)$$

Combining (5.19), (5.20), and (5.23) along with the a-g inequality yields the desired inequality with

$$\alpha_1 = 4.633T^2, \qquad \alpha_2 = 3T, \qquad \alpha_3 = 3[c_2(R,T)]^2 .$$

THE ROBIN PROBLEM

Theorem 5.4. Let u be piecewise C^2 in x, piecewise C^1 in t throughout D, and otherwise arbitrary. Then

$$\int_D u^2 dV \le \alpha_1 \int_D (Lu)^2 dV + \alpha_2 \int_B u^2 dx + \alpha_3 \int_S (\partial u/\partial n + \beta u)^2 d\sigma$$

where $\beta = \beta(x,t)$ is positive on S and α_1, α_2, α_3 are explicitly determined constants [41].

Proof: Using the divergence theorem and Green's second identity we can write

$$\int_D u^2 dV = -\int_D (T-t)_{,t}\, u^2 dV$$

$$= T\int_B u^2 dx + 2\int_D (T-t)\, u\, \{\Delta u - Lu\}\, dV$$

$$= T\int_B u^2 dx + 2\int_S (T-t)\, u\, (\partial u/\partial n + \beta u)\, d\sigma \tag{5.24}$$

$$- 2\int_D (T-t)\, u_{,i} u_{,i}\, dV - 2\int_S (T-t)\, \beta u^2 d\sigma$$

$$- 2\int_D (T-t)\, u\, Lu\, dV \quad .$$

An application of the a-g inequality yields the inequality

$$\int_D u^2 dV \le T\int_B u^2 dx + \frac{1}{4}\int_S (T-t)\, \beta^{-1} (\partial u/\partial n + \beta u)^2 d\sigma - 2\int_D (T-t) u\, Lu\, dV$$

where we have also dropped the negative third term in (5.24). A second application of the a-g inequality then gives

$$\int_D u^2 dV \le (1-\alpha)^{-1} \{T\int_B u^2 dx + \frac{1}{4}\int_S (T-t)\, \beta^{-1} (\partial u/\partial n + \beta u)^2 d\sigma$$

$$+ \alpha^{-1}\int_D (T-t)^2\, (Lu)^2 dV\}$$

for positive $\alpha < 1$.

Starting with (5.24) we can obtain the following result for a semi-linear equation:

Theorem 5.5. Let v and φ be functions defined in D, piecewise C^2 with respect to x and piecewise C^1 with respect to t. Let v satisfy the semi-linear equation

$$Lv = f(x,t,v) \ ,$$

where f satisfies the Lipschitz condition

$$|f(x,t,u_1) - f(x,t,u_2)| \leqq M\,|u_1 - u_2|$$

uniformly in the set $(x,t) \in \bar{D}$, $-\infty < u < \infty$. Then, setting $\psi = v - \varphi$,

$$\int_D \psi^2 dV \leqq \alpha_1 \int_B \psi^2 \, dx + \alpha_2 \int_S \left(\frac{\partial\psi}{\partial n} + \beta\psi\right)^2 d\sigma + \alpha_3 \int_D F^2 dV \ ,$$

where $\beta = \beta(x,t)$ is a bounded, strictly positive, piecewise continuous function,

$$F(x,t) = f(x,t,\varphi) - L(\varphi) \ ,$$

and the α_1, α_2, α_3 are explicitly determined constants.

Proof: Using the a-g inequality we obtain

$$\begin{aligned} \int_D u^2 dV \leq T \int_B u^2 dx + \frac{1}{4} \int_S (T-t)\, \beta^{-1} (\partial u/\partial n + \beta u)^2 d\sigma \\ - 2\int_D (T-t)\, u,_i u,_i dV - 2\int_D (T-t)\, u\, Lu\, dV \ . \end{aligned} \tag{5.25}$$

Adding and subtracting $2\int (T-t)\, bu^2 dV$, b a positive constant, on the right side of (5.25) and using the a-g inequality we have

$$\begin{aligned} \int_D u^2 dV \leq T \int_B u^2 dx + \frac{1}{4} \int_S (T-t)\, \beta^{-1} (\partial u/\partial n + \beta u)^2 d\sigma \\ - 2\int_D (T-t)\, u,_i u,_i dV + \int_D (T-t)^2\, u^2 dV \\ + \int_D (T-t)(Lu - bu)^2 dV - 2\int_D (T-t)\, bu^2 dV \ . \end{aligned} \tag{5.26}$$

Now let u be such that $u = \psi\, e^{-b(T-t)}$, then

$$L\psi = (Lu-bu)\, e^{-b(T-t)}$$

and

$$|L\psi| \leq M\, |\psi| + |F| \; .$$

Thus

$$\int_D (T-t)(Lu-bu)^2\, dV = \int_D (T-t)\, e^{2b(T-t)} (L\psi)^2\, dV$$

$$\leq 2M^2 \int_D (T-t)\, u^2\, dV + 2 \int_D (T-t)\, e^{2b(T-t)} F^2\, dV \qquad (5.27)$$

and if we choose $b = \frac{1}{2}\,[1+2M^2]$ and use (5.27) in (5.26) we obtain

$$\int_D u^2\, dV \leq T \int_B u^2\, dx + \frac{1}{4} \int_S (T-t)\, \beta^{-1}\, (\partial u/\partial n+\beta u)^2\, d\sigma$$

$$+ 2 \int_D (T-t)\, e^{2b(T-t)}\, F^2\, dV \; .$$

6 *A priori* inequalities III – Pseudoparabolic applications

Applications involving pseudoparabolic equations are, for the most part, of quite recent origin. Nevertheless they are appearing with more and more frequency and have by now assumed an important role in a variety of physical situations. Among these we mention nonsteady flows of second order fluids [48]; the seepage of homogeneous fluids through fissured rock [1]; the diffusion of "imprisoned" resonant radiation through a gas [21], [23], [46] (which has applications in the analysis of certain laser systems [35]) and the two-temperature theory of heat conduction [9].

The a priori inequalities developed in this chapter have applications in the boundary value problems which describe the physical problems mentioned above. Because of the relationship between the pseudoparabolic operator treated here,

$$Lu \equiv \Delta u + \Delta u_t - u_t ,$$

and the heat operator, the results of this chapter and their method of proof, bear a close resemblance to that of Chapter 5. For this reason we will abbreviate the proofs somewhat.

INEQUALITIES FOR THE DIRICHLET PROBLEM

Theorem 6.1. Let u be piecewise $C^3(D)$ in x, piecewise $C^1(D)$ in t but otherwise arbitrary. Then

$$\int_D u^2 dV \le \alpha_1 \int_D (Lu)^2 dV + \alpha_2 \int_B u^2 dx + \alpha_3 \int_S u^2 d\sigma$$

where α_1, α_2, α_3 are constants to be determined [44].

Proof: Much of the proof in this case closely parallels that of Theorem 5.1 and so we shall go over the proof fairly rapidly without omitting essential details.

We introduce the auxiliary function w which satisfies

$$
\begin{aligned}
L^{*}w \equiv \Delta(w-w_t) + w_t &= u \quad \text{in} \quad D \cup B \\
w &= 0 \quad \text{on} \quad B_T \cup \bar{S}
\end{aligned}
\tag{6.1}
$$

and then apply the divergence theorem to obtain

$$\int_D u^2 dV = \int_S (u+u_t)\, \partial w/\partial n \, d\sigma + \int_D w\, Lu\, dV + \int_B (\Delta w - w)\, u\, dx$$

from which follows

$$
\begin{aligned}
\int_D u^2 dV \le \Big\{ \int_S (u^2+u_t^2)\, d\sigma \int_S (\partial w/\partial n)^2 d\sigma \Big\}^{\frac{1}{2}} + \Big\{ \int_B [w^2+(\Delta w)^2]\, dx + \lambda_1 \int_D w^2 dV \Big\}^{\frac{1}{2}} \\
\cdot \Big\{2 \int_B u^2 dx + \lambda_1^{-1} \int_D (Lu)^2 dV \Big\}^{\frac{1}{2}}
\end{aligned}
\tag{6.2}
$$

As before the object now is to bound the integrals containing w on the right-hand side of (6.2) by $\int_D u^2 dV$.

To obtain the desired bound on $\int_B [w^2+(\Delta w)^2]\, dx + \lambda_1 \int_D w^2 dV$ we proceed as follows:

$$
\begin{aligned}
-\int_B [w^2+(\Delta w)^2]\, dx &= \int_D (\partial[w^2+(\Delta w)^2]/\partial t)\, dV \\
&= 2\int_D (w-\Delta w)\, L^{*}w\, dV + 2\int_D w[\Delta(w_t - w)]dV \\
&\quad + 2\int_D \Delta w(\Delta w + w_t)\, dV \quad .
\end{aligned}
\tag{6.3}
$$

Now using (3.2) we have that

$$\int_B [w^2+(\Delta w)^2]dx + \lambda_1 \int_D w^2 dV \le \int_B [w^2+(\Delta w)^2]dx + 2\int_D w_{,i}w_{,i}dV - \lambda_1 \int_D w^2 dV \tag{6.4}$$

and substitution of this inequality into (6.3) yields

$$\begin{aligned}\int_B [w^2+(\Delta w)^2]dx + \lambda_1 \int_D w^2 dV \le &- 2\int_D (w-\Delta w)\, L^*w\, dV - 2\int_D w[\Delta(w_t-w)]dV \\ &- 2\int_D \Delta w(\Delta w+w_t)dV + 2\int_D w_{,i}w_{,i}dV - \lambda_1 \int_D w^2 dV.\end{aligned} \tag{6.5}$$

By the divergence theorem and the initial-boundary conditions satisfied by w we have

$$- 2\int_D (\Delta w)\, w_t dV = - 2\int_D w\,\Delta\, w_t dV = -\int_B w_{,i}w_{,i}dx \le 0 . \tag{6.6}$$

Thus, if in (6.5) we drop the negative terms indicated by (6.6), recall that $\int_D w\,\Delta w\, dV = -\int_D w_{,i}w_{,i}dV$, and use an appropriately weighted a-g inequality we obtain

$$\int_B [w^2+(\Delta w)^2]dx + \lambda_1 \int_D w^2 dV \le (2\lambda_1)^{-1} \cdot (2+\lambda_1)\int_D u^2 dV \tag{6.7}$$

Before proceeding to bound the second term on the right-hand side of (6.2), we derive three inequalities which will be useful later. We start off with

$$\int_D (\Delta w)^2 dV = \int_D \Delta w\, L^*w\, dV + 2\int_D \Delta w\, \Delta w_t dV - \int_D w_t\, \Delta w\, dV$$

which, by the divergence theorem and (6.1) gives

$$\begin{aligned}\int_D (\Delta w)^2 dV &= \int_D \Delta w\, L^*w\, dV - \frac{1}{2}\int_B (\Delta w)^2 dx - \frac{1}{2}\int_B w_{,i}w_{,i}\, dx \\ &\le \int_D \Delta w\, L^*w\, dV .\end{aligned} \tag{6.8}$$

The Schwarz inequality now yields

$$\int_D (\Delta w)^2 dV \le \int_D (L^* w)^2 dV = \int_D u^2 dV \quad . \tag{6.9}$$

The second inequality we need follows from (6.8) also, for we have

$$\frac{1}{2} \int_B (\Delta w)^2 dx \le \int_D \Delta w \; L^* w \; dV$$

which yields, by the Schwarz inequality and (6.9)

$$\frac{1}{2} \int_B (\Delta w)^2 dx \le \int_D u^2 dV \quad . \tag{6.10}$$

To obtain the third inequality we use the fact that $w = 0$ on S to write

$$\int_D w,_i w,_i dV = - \int_D w\Delta w \; dV = - \int_D w \; L^* w \; dV - \int_D w\Delta w_t dV + \int_D w \; w_t dV \; . \tag{6.11}$$

But by (6.6) and since $\int_D w \; w_t dV = - \frac{1}{2} \int_B w^2 dx \le 0$, we have $\int_D w,_i w,_i dV \le$ $-\int_D w \; L^* w \; dV$, which by (3.2) and the Schwarz inequality, yields

$$\int_D w,_i w,_i dV \le \lambda_1^{-1} \int_D u^2 dV \; . \tag{6.12}$$

To bound the second term on the right-hand side of (6.2) we start with the identity (2.8) where the tangential derivative terms drop out since $w = 0$ on S:

$$\begin{aligned} \int_S f^i n_i (\partial w/\partial n)^2 d\sigma &= 2 \int_D f^\alpha w,_\alpha L^* w \; dV - \int_D \{f^\alpha_{,\alpha} w,_i w,_i - 2 f^j_{,i} w,_i w,_j\} dV \\ &+ 2 \int_D f^{N+1}_{,i} \; w_t w,_i dV - 2 \int_D f^\alpha w,_\alpha (w_t - \Delta w_t) dV \\ &- 2 \int_S f^{N+1} \; w_t \; \partial w/\partial n \; d\sigma + \int_{B \cup B_T} f^\alpha n_\alpha \; w,_i w,_i \; dx \quad . \end{aligned} \tag{6.13}$$

Some terms drop immediately; for instance since $w = 0$ on S and w_t is a derivative in a tangential direction on S, the second from the last term drops. For the same reason $\int_{B_T} f^{\alpha} n_{\alpha} w_{,i} w_{,i} dx = 0$ and on B, $n_i = 0$, $i=1,2,\cdots,N$ and $n_{N+1} = -1$ so that this term reduces to $-\int_B f^{N+1} w_{,i} w_{,i} dx \le 0$ if $f^{N+1} \ge 0$, a condition we now impose. Further, consider some of the terms in $-2\int_D f^{\alpha} w_{,\alpha}(w_t - \Delta w_t) dV$. We have

$$\int_D f^{N+1} w_t \, \Delta w_t dV = -\int_D f^{N+1} w_{,it} w_{,it}, \quad \text{if} \quad f^{N+1}_{,i} = 0, \qquad i=1,2,\cdots,N$$

and

$$\int_D f^i w_{,i} \Delta w_t dV = -\int_B f^i w_{,i} \Delta w \, dx - \int_D f^i w_{,it} \Delta w \, dV, \quad \text{if} \quad f^i_t = 0,$$

$$i=1,2,\cdots,N$$

Using these in (6.13) along with the a-g inequality we have

$$\begin{aligned}
\int_S f^i n_i (\partial w/\partial n)^2 d\sigma \le\ & 2\int_D f^{N+1}(L^* w)^2 dV + \frac{1}{2}\int_D f^{N+1}(w_t)^2 dV + \int_D f^i w_{,i} L^* \, dV \\
& + C\int_D w_{,i} w_{,i} dV + \int_D \beta^{-1} f^{N+1}_{,i} f^{N+1}_{,i} (w_t)^2 dV + \int_D \beta \, w_{,i} w_{,i} dV \\
& - 2\int_D f^{N+1}(w_t)^2 dV + 2\int_D \frac{f^j f^j}{f^{N+1}} w_{,i} w_{,i} dV + \frac{1}{2}\int_D f^{N+1}(w_t)^2 dV \\
& -2\int_D f^{N+1} w_{,it} w_{,it} dV - 2\int_B f^i w_{,i} \Delta w \, dx - 2\int_D f^i w_{,it} \Delta w \, dV \\
& - \int_B f^{N+1} w_{,i} w_{,i} dx ,
\end{aligned}$$

where we have further imposed the condition that $f^{N+1} \ge$ in $\bar{D}$. If we now choose $\beta = f^{N+1}_{,i} f^{N+1}_{,i} / f^{N+1}$ we cancel out the terms containing $(w_t)^2$ and dropping all negative terms we obtain

$$\begin{aligned}
\int_S f^i n_i (\partial w/\partial n)^2 d\sigma \le\ & 2\,|f^{N+1}|_M \int_D (L^* w)^2 dV + \int_D f^i w_{,i} L^* w \, dV \\
& + \left\{ \left| \frac{f^{N+1}_{,j} f^{N+1}_{,j} + 2 f^j f^j}{f^{N+1}} \right|_M + C \right\} \int_D w_{,i} w_{,i} dV - \int_B f^{N+1} w_{,i} w_{,i} dx \qquad (6.14) \\
& - 2\int_D f^{N+1} w_{,it} w_{,it} dV - 2\int_B f^i w_{,i} \Delta w \, dx - 2\int_D f^i w_{,it} \Delta w \, dV \ .
\end{aligned}$$

Now, straightforward applications of the arithmetic-geometric and Schwarz inequalities result in the following:

$$- 2 \int_B f^i w,_i \Delta w \, dx \le \int_B f^{N+1} w,_i w,_i \, dx + \int_B \frac{f^i f^i}{f^{N+1}} (\Delta w)^2 dx$$

$$- 2 \int_D f^i w,_{it} \Delta w \, dV \le 2 \int_D f^{N+1} w,_{it} w,_{it} dV + \frac{1}{2} \int_D \frac{f^i f^i}{f^{N+1}} (\Delta w)^2 dV$$

$$\int_D f^i w,_i L^* w \, dV \le \frac{|f^\ell f^\ell|_M}{2} \int_D w,_i w,_i dV + \frac{1}{2} \int_D (L^* w)^2 \, dV \quad .$$

Combining these inequalities with (6.14) and using (6.9), (6.10) and (6.12) yields

$$\begin{aligned} \int_S f^i n_i (\partial w/\partial n)^2 d\sigma \le \Bigg\{ \lambda_1^{-1} \left| \frac{f_{,i}^{N+1} f_{,i}^{N+1} + f^i f^i}{f^{N+1}} \right|_M + \frac{|f^i f^i|_M}{2} + C \\ + \frac{3}{2} \left| \frac{f^i f^i}{f^{N+1}} \right|_M + 2 \, |f^{N+1}|_M + \frac{1}{2} \Bigg\} \int_D u^2 dV \equiv \hat{K} \int_D u^2 dV \end{aligned} \tag{6.15}$$

so that

$$\int_S (\partial w/\partial n)^2 d\sigma \le K \int_D u^2 dV$$

where $K = \hat{K}/p_m$. Thus the theorem holds for $\alpha_1 = (\lambda_1 + 2) \lambda_1^{-2}$, $\alpha_2 = 2(1 + 2\lambda_1^{-1})$, $\alpha_3 = 2K$.

AN INEQUALITY FOR A SEMI-LINEAR PROBLEM

Applying techniques similar to those used in the preceding section we can derive bounds which apply to the Dirichlet problem when the pseudoparabolic equation is semi-linear. This is done in the following theorem.

Theorem 6.2. Let v be a solution of the semi-linear equation $Lv = f(x,t,v)$ where f satisfies a Lipschitz condition in v with Lipschitz constant M. Let $\psi = v - \varphi$ where φ is an arbitrary $C^3(D)$ function. Then

$$\int_D \psi^2 \, dV \leq \alpha_1 \int_B \psi^2 \, dx + \alpha_2 \int_S (\psi^2+\psi_t^2) \, d\sigma + \alpha_3 \int_D F^2 dV ,$$

where $F(x,t) = f(x,t,\varphi) - L(\varphi)$ and α_1, α_2, α_3 are explicitly determined constants depending on domain geometry, M, and a positive constant b to be introduced shortly.

Proof: The derivation follows closely that of the preceding section after a few preliminaries. Introduce the function

$$u = \psi \, e^{b(\tau-t)} ,$$

where b is a positive constant. Then

$$L(\psi) = (Lu-bu) \, e^{-b(\tau-t)} , \tag{6.16}$$

and we now want to bound

$$\int_D e^{2b(\tau-t)} \psi^2 dV = \int_D u^2 dV .$$

To do this introduce the function w which satisfies the initial-boundary value problem

$$L^* w - bw = u \quad \text{in} \quad D ,$$

$$w = 0 \quad \text{on} \quad S \cup B_T .$$

Proceeding as in the previous section we obtain

$$\begin{aligned} \int_D u^2 dV \leq \{ \int_S (u^2+u_t^2) \, d\sigma \int_S (\partial w/\partial n)^2 \}^{\frac{1}{2}} + \{ \int_B [w^2+(\Delta w)^2] \, dx \\ + (b+\lambda_1) \int_D w^2 \, dV\}^{\frac{1}{2}} \{2 \int_B u^2 dx + \lambda_1^{-1} \int_D (Lu-bu)^2 dV\}^{\frac{1}{2}} \end{aligned} \tag{6.17}$$

and the two important inequalities corresponding to (6.7) and (6.15) are

$$\int_B [w^2+(\Delta w)^2] \, dx + (b+\lambda_1) \int_D w^2 dV \leq \left\{\frac{2+b+\lambda_1}{2(b+\lambda_1)}\right\} \int_D u^2 dV \tag{6.18}$$

and

$$\int_S (\partial w/\partial n)^2 d\sigma \le K' \int_D u^2 dV , \tag{6.19}$$

where $K' = K + b|f^{\alpha}_{,\alpha}|_M/2\lambda_1^2$.

To complete the bound we write

$$|L(\psi)| \le M|\psi| + F \tag{6.20}$$

so that

$$\int_D (Lu-bu)^2 dV \le 2M^2 \int_D e^{2b(\tau-t)}\psi^2 dV + 2\int_D e^{2b(\tau-t)}F^2 dV$$

by (6.15) and (6.20). The bound then follows from (6.17), (6.18), (6.19) and (6.21), and we obtain

$$\alpha_1 = \frac{2(2+b+\lambda_1)}{\bar{K}^2 \cdot (b+\lambda_1)} ,$$

$$\alpha_2 = 2K'/\bar{K}^2 ,$$

$$\alpha_3 = \frac{2(2+b+\lambda_1)\ e^{2b\tau}}{\bar{K}^2 \cdot (b+\lambda_1)} ,$$

where $\bar{K} = 1 - M(2+b+\lambda_1)^{\frac{1}{2}}/(b+\lambda_1)$ and b is chosen so that $\bar{K} > 0$.

BOUNDS FOR THE NEUMANN PROBLEM

The appropriate inequality for applications to the Neumann problem is given in Theorem 6.3 below. The proof of the theorem has as its starting point the identity

$$\int_{B_t} (u^2+u_{,i}u_{,i})dx = \int_B (u^2+u_{,i}u_{,i})dx - 2\int_0^t \int_{B_t} u\ L\ u\ dxd\tau$$
$$- 2\int_0^t \int_{B_\tau} u_{,i}u_{,i} dxd\tau + 2\int_0^t \int_{S_\tau} u \frac{\partial}{\partial n}(u+u_t)\ dsd\tau$$

and then exactly parallels the proof of Theorem 5.3 and thus we omit the details.

Theorem 6.3. Let u be an arbitrary piecewise $C^3(D)$ function. Then

$$\int_D u^2 dV \le \alpha_1 \int_D (Lu)^2 dV + \alpha_2 \int_B (u^2 + u,_i u,_i) dx + \alpha_3 \int_S (\partial u/\partial n)^2 d\sigma$$

where α_1, α_2 and α_3 are explicitly determined constants [19].

AN INEQUALITY FOR A BOUNDARY VALUE PROBLEM OF MILNE

We now develop an a priori inequality which has applications to a pseudo-parabolic initial-boundary value problem which is not related to an analogous parabolic problem as in the case of the Dirichlet and Neumann problems. This initial-boundary value problem is due to Milne [21], [23], and originated in his study of the diffusion of resonance radiation through a gas. It has recently found applications in the analysis of optically pumped lasers [35]. The problem is given by

$$Lu = f(x,t) \quad \text{in} \quad D,$$

$$u = g(x,0) \quad \text{on} \quad B,$$

$$u + u_t + \frac{\partial}{\partial n}(u + u_t) = h(x,t) \quad \text{on} \quad S.$$

Hereafter we shall refer to this as the Milne problem.

The appropriate a priori inequality for application to the Milne problem is given in Theorem 6.4.

Theorem 6.4. Let u be an arbitrary piecewise C^3 function in D. Then

$$\int_D u^2 dV \le \alpha_1 \int_D (Lu)^2 dV + \alpha_2 \int_B (u^2 + u,_i u,_i) dx + \alpha_3 \int_S (u + u_t + \partial(u + u_t)/\partial n)^2 d\sigma$$

where α_1, α_2 and α_3 are explicitly determined constants.

Proof: The proof follows closely that of Theorem 5.4 of the preceding theorem. We quickly sketch the proof using a different weighting function. Realizing that

$$\int_D e^{(T-t)} u^2 dV = -\int_D [e^{(T-t)}]_t u^2 dV$$

and using the divergence theorem we have

$$\int_D e^{(T-t)} u^2 dV = e^T \int_B u^2 dx + 2\int_D e^{(T-t)} uu_t \, dV .$$

Upon using $u_t = (u+u_t),_{ii} - Lu$ and the divergence theorem, we obtain

$$\int_D e^{(T-t)} u^2 dV = e^T \int_B u^2 dx + 2\int_S e^{(T-t)} u\left\{\frac{\partial}{\partial n}(u+u_t) + (u+u_t)\right\} d\sigma$$

$$- 2\int_D e^{(T-t)} u,_i (u+u_t),_i dV - 2\int_D e^{(t-t)} u \, Lu \, dV \qquad (6.21)$$

$$- 2\int_S e^{(T-t)} u^2 d\sigma - 2\int_S e^{(T-t)} uu_t \, d\sigma$$

Now apply the divergence theorem to $\int_D e^{(T-t)} u,_i u_{t,i} dV$ and $\int_S e^{(T-t)} uu_t \, d\sigma$ to obtain

$$2\int_D e^{(T-t)} u,_i u_{t,i} dV = - e^T \int_B u,_i u,_i dx + \int_D e^{(T-t)} u,_i u,_i dV$$

and

$$2\int_S e^{(T-t)} uu_t d\sigma = - e^T \oint_{\partial B} u^2 dS + \int_S e^{(T-t)} u^2 d\sigma \quad .$$

Using these in (6.21) and dropping nonpositive terms we have

$$\int_D e^{(T-t)} u^2 dV \le e^T \int_B (u^2 + u,_i u,_i) dx + e^T \oint_{\partial B} u^2 dS$$

$$+ 2\int_S e^{(T-t)} u\left\{\frac{\partial}{\partial n}(u+u_t) + u+u_t\right\} d\sigma$$

$$- 2\int_D e^{(T-t)} u \, Lu \, dV - 2\int_S e^{(T-t)} u^2 d\sigma$$

and an application of the a-g inequality yields

$$\int_D e^{(T-t)} u^2 dV \le 2\, e^T \int_B (u^2+u_{,i}u_{,i})dx + 2\, e^T \oint_{\partial B} u^2 dS$$

$$+\int_S e^{(T-t)} \left\{\frac{\partial}{\partial n} (u+u_t) + u + u_t\right\}^2 d\sigma + 4 \int_D e^{(T-t)} (Lu)^2 dV .$$

7 *A priori* inequalities IV – Fourth order elliptic applications

This chapter is concerned with explicit a priori inequalities which have applications to fourth order elliptic boundary value problems. Although we will emphasize results for the biharmonic operator, Δ^2, explicit a priori inequalities exist which apply to more general fourth order elliptic operators (in particular see [38] and [42]).

AN INEQUALITY FOR THE FIRST BIHARMONIC BOUNDARY VALUE PROBLEM

The first biharmonic boundary value problem is given by

$$\begin{aligned} \Delta^2 u = f \quad & \text{in} \quad B, \\ u = g \quad & \text{on} \quad \partial B \\ \partial u/\partial n = h \quad & \text{on} \quad \partial B, \end{aligned}$$

where f, g and h are known prescribed functions. An inequality with applications to this problem is given by the following.

Theorem 7.1. Let u be any piecewise C^4 function in $\bar{B}$. Then

$$\int_B u^2 dx \le \alpha_1 \int_B (\Delta^2 u)^2 dx + \alpha_2 \oint_{\partial B} u^2 dS + \alpha_3 \oint_{\partial B} (\partial u/\partial n)^2 dS + \alpha_4 \oint_{\partial B} (\partial u/\partial s)^2 dS$$

where, as usual, α_1, α_2, α_3 and α_4 are explicitly determined constants.

Proof: Introduce the function φ which satisfies

$$\begin{aligned} \Delta\varphi = 0 \quad & \text{in} \quad B, \\ \varphi = u \quad & \text{on} \quad \partial B \quad . \end{aligned} \tag{7.1}$$

Let $v = u - \varphi$ so that

$$\Delta v = \Delta u \quad \text{in} \quad B,$$
$$v = 0 \quad \text{on} \quad \partial B$$

and thus by the triangle inequality we have

$$\left(\int_B u^2 dx\right)^{\frac{1}{2}} \leq \left(\int_B v^2 dx\right)^{\frac{1}{2}} + \left(\int_B \varphi^2 dx\right)^{\frac{1}{2}} \quad . \tag{7.2}$$

The object now is to bound the two integrals on the right-hand side in terms of known quantities. To do this for $\int_B v^2 dx$ we introduce the function w satisfying

$$\begin{aligned} \Delta^2 w &= v \quad && \text{in} \quad B\ , \\ w = \partial w/\partial n &= 0 \quad && \text{on} \quad \partial B\ . \end{aligned} \tag{7.3}$$

Then since both v, w and $\partial w/\partial n$ vanish on ∂B we have

$$\int_B (v\ \Delta^2 w - w\ \Delta^2 v)dx = - \oint_{\partial B} \Delta(w)\ \partial v/\partial n\ dS$$

so that

$$\int_B v^2 dx = \int_B w\ \Delta^2 u\ dx - \oint_{\partial B} \Delta(w)\ \partial v/\partial n\ dS \quad ,$$

and the Schwarz inequality yields

$$\int_B v^2 dx \leq \left(\int_B w^2 dx \int_B (\Delta^2 u)^2 dx\right)^{\frac{1}{2}} + \left(\oint_{\partial B} (\Delta w)^2 dS \oint_{\partial B} (\partial v/\partial n)^2 dS\right)^{\frac{1}{2}} \tag{7.4}$$

We now bound $\int_B w^2 dx$ and $\oint_{\partial B} (\Delta w)^2 dS$ in terms of $\int_B v^2 dx$. By (3.6) we have

$$\int_B w^2 dx \leq \lambda_1^{-2} \int_B (\Delta w)^2 dx = \lambda_1^{-2} \int_B w\ \Delta^2 w\ dx \tag{7.5}$$

and using the Schwarz inequality

$$\int_B w^2 dx \le \lambda_1^{-4} \int_B (\Delta^2 w)^2 dx = \lambda_1^{-4} \int_B v^2 dx \quad . \tag{7.6}$$

Furthermore

$$\int_B w_{,i} w_{,i} dx = \int_B w \, \Delta w \, dx \le \left(\int_B w^2 dx\right)^{\frac{1}{2}} \left(\int_B (\Delta w)^2\right)^{\frac{1}{2}}$$

$$\le \left(\lambda_1^{-1} \int_B w_{,i} w_{,i} dx\right)^{\frac{1}{2}} \left(\int_B (\Delta w)^2 dx\right)^{\frac{1}{2}}$$

so that

$$\int_B w_{,i} w_{,i} dx \le \lambda_1^{-1} \int_B (\Delta w)^2 dx = \lambda_1^{-1} \int_B w \, \Delta^2 w \, dx \; . \tag{7.7}$$

The a-g inequality yields

$$\int_B w \, \Delta^2 w \, dx \le \frac{\alpha}{2} \int_B w^2 dx + \frac{1}{2\alpha} \int_B (\Delta^2 w)^2 dx \tag{7.8}$$

so that if we use (7.6) and (7.8) in (7.7) we obtain

$$\int_B w_{,i} w_{,i} dx \le \lambda_1^{-2} \int_B (\Delta^2 w)^2 dx = \lambda_1^{-2} \int_B v^2 dx \tag{7.9}$$

choosing $\alpha = \lambda_1^2$. Further (7.5) and (7.6) along with the Schwarz inequality imply that

$$\int_B (\Delta w)^2 dx \le \lambda_1^{-2} \int_B v^2 dx \; . \tag{7.10}$$

Inequalities (7.9) and (7.10) will be useful shortly. One more inequality along these lines is needed. Since $w = \partial w/\partial n = 0$ on ∂B it follows that

$$\int_B (\Delta w)^2 dx = \int_B w \, \Delta^2 w \, dx = \int_B w_{,ij} w_{,ij} dx$$

so that

$$\int_B w_{,ij}w_{,ij}dx \le \lambda_1^{-2} \int_B v^2 dx \ . \tag{7.11}$$

We are now ready to bound $\oint_{\partial B} (\Delta w)^2 dS$. Denote by f^i, $i=1,\cdots,N$, the ith component of a vector field with piecewise continuous second derivatives in B and apply the Green first identity to $\int_B f^\ell w_{,\ell}\, \Delta^2 w\, dx$ to obtain

$$\int_B f^\ell w_{,\ell}\, \Delta^2 w\, dx = \int_B \Delta(f^\ell w_{,\ell})\, \Delta w\, dx - \oint_{\partial B} \Delta w \frac{\partial}{\partial n} (f^\ell w_{,\ell})\, dS \ . \tag{7.12}$$

Performing the indicated operations in the first term on the right-hand side and using the divergence theorem we have

$$\begin{aligned} \int_B \Delta(f^\ell w_{,\ell})\Delta w\, dx &= \int_B w_{,\ell}\Delta f^\ell \Delta w\, dx + 2\int_B f^\ell_{,i} w_{,i\ell}\Delta w\, dx \\ &\quad + \frac{1}{2}\oint_{\partial B} f^\ell n_\ell (\Delta w)^2 dS - \frac{1}{2}\int_B f^\ell_{,\ell}(\Delta w)^2 dx \end{aligned} \tag{7.13}$$

and combining (7.12) and (7.13) yields

$$\begin{aligned} \int_B f^\ell w_{,\ell}\Delta^2 w\, dx &= \int_B \{[2f^\ell_{,i} w_{,i\ell} + w_{,\ell}\Delta f^\ell]\, \Delta w - \frac{1}{2} f^\ell_{,\ell}(\Delta w)^2\} dx \\ &\quad + \frac{1}{2}\oint_{\partial B} f^\ell n_\ell (\Delta w)^2 dS - \oint_{\partial B} \Delta w \frac{\partial}{\partial n}(f^\ell w_{,\ell}) dS \ . \end{aligned} \tag{7.14}$$

Now since $w = \partial w/\partial n = 0$ on ∂B we have that all first derivatives of w vanish on ∂B and thus

$$\oint_{\partial B} \Delta w \frac{\partial}{\partial n} (f^\ell w_{,\ell}) dS = \oint_{\partial B} (\Delta w)\, f^\ell \frac{\partial w_{,\ell}}{\partial n}\, dS \tag{7.15}$$

Now $\Delta w = w_{,ii} = n_i \dfrac{\partial w_{,\ell}}{\partial n} + n_\ell (n_\ell \dfrac{\partial}{\partial x_i} - n_i \dfrac{\partial}{\partial x_\ell})\, w_{,i}$ so that on ∂B

$$\Delta w = n_i \frac{\partial w_{,i}}{\partial n} \tag{7.16}$$

since the other term is, on ∂B, a tangential derivate of $w,_i$ and hence vanishes. Further

$$n_i \frac{\partial w,_i}{\partial n} = n_i \frac{\partial}{\partial n} \{n_i \frac{\partial w}{\partial n} + n_\ell (n_\ell \frac{\partial}{\partial x_i} - n_i \frac{\partial}{\partial x_\ell}) w\}$$

so that again, on ∂B,

$$n_i \frac{\partial w,_i}{\partial n} = \frac{\partial^2 w}{\partial n^2} \tag{7.17}$$

where we have used the fact that $n_i \frac{\partial n_i}{\partial n} = \frac{1}{2} (n_i n_i),_j n_j = 0$ on ∂B and $n_i n_i = 1$ there. In a similar manner we find that

$$\frac{\partial w,_\ell}{\partial n} = n_\ell \frac{\partial^2 w}{\partial n^2} \tag{7.18}$$

on ∂B. Then (7.16), (7.17) and (7.18) in (7.15) give

$$\oint_{\partial B} \Delta w \frac{\partial}{\partial n} (f^\ell w,_\ell)\, dS = \oint_{\partial B} f^\ell n_\ell (\Delta w)^2\, dS$$

and using this in (7.14) gives the desired identity

$$\begin{aligned} \oint_{\partial B} f^\ell n_\ell (\Delta w)^2 dS = \int_B \{[2f^\ell,_i w,_{i\ell} + w,_\ell \Delta f^\ell]\Delta w - \tfrac{1}{2} f^\ell,_\ell (\Delta w)^2\} dx \\ - \int_B f^\ell w,_\ell \Delta^2 w\, dx \quad . \end{aligned} \tag{7.19}$$

Now choose the f^ℓ such that $f^\ell n_\ell = p \geq p_m > 0$ on ∂B and use the Schwarz inequality along with (7.9), (7.10) and (7.11) to obtain

$$\begin{aligned} \oint_{\partial B} (\Delta w)^2 dS \leq 2p_m^{-1} \{2|f^\ell,_i f^\ell,_i|_M^{\frac{1}{2}} \lambda_1^{-2} + |(\Delta f^\ell)(\Delta f^\ell)|_M^{\frac{1}{2}} \lambda_1^{-2} + \tfrac{1}{2} |f^\ell,_\ell|_M \lambda_1^{-2} \\ + |f^\ell f^\ell|_M^{\frac{1}{2}} \lambda_1^{-1}\} \int_B v^2 dx \equiv K \int_B v^2 dx \quad . \end{aligned} \tag{7.20}$$

We still need to bound integrals involving the function φ. We have

$$\oint_{\partial B} (\partial v/\partial n)^2 dS = \oint_{\partial B} (\partial u/\partial n - \partial\varphi/\partial n)^2 dS \le 2 \oint_{\partial B} (\partial u/\partial n)^2 dS$$
$$+ 2 \oint_{\partial B} (\partial\varphi/\partial n)^2 dS \tag{7.21}$$

A bound on $\oint_{\partial B} (\partial\varphi/\partial n)^2 dS$ is obtained using the identity (2.6):

$$- \oint_{\partial B} [f^i n_i\{(\partial\varphi/\partial s)^2 - (\partial\varphi/\partial n)^2\} - 2f^i s_i \ \partial\varphi/\partial n\cdot\partial\varphi/\partial s]dS$$
$$\le C \oint_{\partial B} \varphi\, \partial\varphi/\partial n \, dS$$

where we have used the fact that $\int_B \varphi,_i\varphi,_i dx = \oint_{\partial B} \varphi\, \partial\varphi/\partial n \, dS$ since $\Delta\varphi = 0$. Now using weighed a-g inequalities with weight $\alpha = 2\, p_m^{-1}$, $\beta = 4\, p_m^{-1}$ we obtain

$$\oint_{\partial B} (\partial\varphi/\partial n)^2 dS \le 2\, p_m^{-2} \{C^2 \oint_{\partial B} \varphi^2 dS + [|f^i n_i|_M \cdot t_m$$
$$+ 4 |f^i s_i|_M^2] \oint_{\partial B} (\partial\varphi/\partial s)^2 dS\} \tag{7.22}$$

Finally, we note that

$$\int_B \varphi^2 \le \tilde{\alpha}_2 \oint_{\partial B} \varphi^2 dS \equiv \tilde{\alpha}_2 \oint_{\partial B} u^2 dS \tag{7.23}$$

by Theorem 3.1, where $\tilde{\alpha}_2$ is given by α_2 on page 22.

The inequality of the theorem is now obtained using (7.2), (7.4), (7.20)-(7.23) and the a-g mean inequality:

$$\int_B u^2 dx \le 4\lambda_1^{-4} \int_B (\Delta^2 u)^2 dx + (2\tilde{\alpha}_2 + 16\, K\, p_m^{-2} C^2) \oint_{\partial B} u^2 dS + 8K \oint_{\partial B} (\partial u/\partial n)^2 dS$$
$$+ 16K\, p_m^{-2}[|f^i n_i|_M\, p_m + 4|f^i s_i|_M^2] \oint_{\partial B} (\partial u/\partial s)^2 dS \quad .$$

INEQUALITIES FOR OTHER BOUNDARY VALUE PROBLEMS

When we attempt to treat other boundary value problems for the biharmonic operator we immediately run into difficulties. First, explicit a priori inequalities, as we have come to think of them, have not been developed for some of the problems. Second, for those that have, rather lengthy and involved derivations are required. For these reasons we shall summarize the known results and refer the interested reader to the appropriate literature for the details.

Consider then the second boundary value problem for elastic plates:

$$\begin{aligned} \Delta^2 u &= f \quad \text{in} \quad B, \\ u &= g \quad \text{on} \quad \partial B, \\ M(u) &= h \quad \text{on} \quad \partial B, \end{aligned} \tag{7.24}$$

where f,g and h are prescribed data and M(u) is proportional to the normal moment, i.e.,

$$M(u) = \Delta u - (1-\sigma)\left(\frac{\partial^2 u}{\partial s^2} + \frac{1}{\rho}\frac{\partial u}{\partial n}\right).$$

In the above equation σ denotes Poisson's ratio and ρ denotes the radius of curvature on ∂B. Explicit a priori inequalities (in our sense of the term) which have applicability to this problem have not appeared in the literature to our knowledge. However, Bramble and Payne [5] have given the following result:

For any biharmonic function V with piecewise fourth derivatives in a domain B whose boundary has bounded curvature, the following a priori bound holds for any p-th order derivative of V at a point O in B:

$$|V^{(p)}(0)|^2 \le K_1 \oint_{\partial B} V^2\,dS + K_2 \oint_{\partial B} (\partial V/\partial s)^2\,dS + K_3 \oint_{\partial B} (\partial^2 V/\partial s^2)^2\,dS + K_4 \oint_{\partial B} [M(V)]^2\,dS$$

where the constants K_1, K_2, K_3, K_4 (which depend on p) are explicitly determined.

They later indicate how the above inequality leads to pointwise bounds for a solution u of the boundary value problem (7.24).

The third boundary value problem for elastic plates is:

$$\Delta^2 u = f \quad \text{in} \quad B,$$

$$M(u) = g \quad \text{on} \quad \partial B,$$

$$Q(u) = h \quad \text{on} \quad \partial B.$$

Here Q(u) is proportional to the reaction normal to the plate, i.e.,

$$Q(u) = \frac{\partial(\Delta u)}{\partial n} - (1-\sigma)\left[\frac{\partial^3 u}{\partial s^2 \partial n} - \frac{\partial}{\partial s}\left(\frac{1}{\rho}\frac{\partial u}{\partial s}\right)\right] .$$

The following result [5] is applicable to this problem.

> Let V be any function with piecewise fourth derivatives in B which satisfies the conditions
>
> $$\oint_{\partial B} V dS = \oint_{\partial B} (\partial V/\partial x) dS = \oint_{\partial B} (\partial V/\partial y) dS = 0 .$$
>
> Then at any point $0 \in B$,
>
> $$|V^{(p)}(0)|^2 \le k_1 \oint_{\partial B} [M(V)]^2 dS + k_2 \oint_{\partial B} [Q(V)]^2 dS + k_3 \int_B \eta (\Delta^2 V)^2 dxdy$$
>
> where the constants k_1, k_2, k_3, k_4 are explicitly determined, $p \le 3$, and η is a known nonnegative function in B.

Explicit a priori inequalities of the type discussed in this volume have been given by Bramble and Payne [8] for two of the following three mixed boundary value plate problems for elastic plates:

Problem I:

$\Delta^2 u$ prescribed in B,

$u, \dfrac{\partial u}{\partial n}$ given on ∂B_1,

$u, M(u)$ given on ∂B_2;

Problem II:

$\Delta^2 u$ prescribed in B,

$u, \dfrac{\partial u}{\partial n}$ given on ∂B_1,

$M(u)$, $V(u)$ given on ∂B_2;

Problem III:

$\Delta^2 u$ prescribed in B,

$u, M(u)$ given on ∂B_1

$M(u)$, $V(u)$ given on ∂B_2.

In the above it is assumed that the boundary ∂B of B consists of two disjoint connected portions ∂B_1 and ∂B_2 on which the different sets of boundary conditions are imposed.

A priori inequalities with applications to these boundary value problems are:

for Problem I:

$$\int_B w^2 \, dxdy \le \alpha_1 \int_B (\Delta^2 w)^2 \, dxdy + \alpha_2 \oint_{\partial B} w^2 \, dS + \alpha_3 \oint_{\partial B} (\partial w/\partial s)^2 \, dS + \alpha_4 \oint_{\partial B} (\partial^2 w/\partial s^2)^2 \, dS$$

$$+ \alpha_5 \oint_{\partial B_1} (\partial w/\partial n)^2 \, dS + \alpha_6 \oint_{\partial B_1} (\partial^2 w/\partial s \partial n)^2 \, dS + \alpha_7 \oint_{\partial B_2} [M(w)]^2 \, dS;$$

for Problem II:

$$\int_B w^2 \, dxdy \le \beta_1 \int_B (\Delta^2 w)^2 \, dxdy + \beta_2 \oint_{\partial B_1} w^2 \, dS + \beta_3 \oint_{\partial B_1} (\partial w/\partial s)^2 \, dS + \beta_4 \oint_{\partial B_1} (\partial w/\partial n)^2 \, dS$$

$$+ \beta_5 \oint_{\partial B_1} (\partial^2 w/\partial s^2)^2 \, dS + \beta_6 \oint_{\partial B_1} (\partial^2 w/\partial s \partial n)^2 \, dS + \beta_7 \oint_{\partial B_2} [M(w)]^2 \, dS$$

$$+ \beta_8 \oint_{\partial B_2} [V(w)]^2 \, dS \; .$$

The constants appearing in these inequalities are, as usual, known explicitly.

Problem III is the most difficult of the three since for certain regions it does not have a unique solution. As Bramble and Payne point out, "Suppose, for instance, that the domain B lies in the right half plane and ∂B_1 is a portion of the y-axis. For this case the plate may rotate as a rigid body about the y-axis. The solution will be unique only up to a rigid body rotation, a term of the form $w = \alpha x$." Because of the inherent difficulty in treating problem III they were unable to obtain an L_2 bound for $\int_B w^2 dxdy$ but instead had to settle for an L_2 bound for the "energy integral"

$$E(w,w) = \int_B [\sigma(\Delta w)^2 + (1-\sigma)\, w,_{ij} w,_{ij}]dxdy$$

so called because E(u,u) is proportional to the stress energy when u is the solution of problem III. Their inequality is

$$E(w,w) \le \gamma_1 \int_B (\Delta^2 w)^2 dxdy + \gamma_2 \oint_{\partial B} [M(w)]^2 dS + \gamma_3 \oint_{\partial B_2} [V(w)]^2 dS$$

$$+ \gamma_4 \oint_{\partial B_1} (\partial w/\partial s)^2 dS + \gamma_5 \oint_{\partial B_1} (\partial^2 w/\partial s^2)^2 dS \quad .$$

Following the techniques of [2], [5], it is possible to obtain pointwise bounds for any second derivatives of w in terms of L_2 integrals of the data.

8 Pointwise bounds

In the preceding chapters we have emphasized the use of explicit _a priori_ inequalities in computing approximate solutions of boundary value problems. We now turn our attention to another aspect of their use, the calculation of pointwise bounds for any sufficiently smooth approximate solution.

The technique of obtaining pointwise bounds is similar whether we are concerned with parabolic or elliptic operators. Both situations require knowledge of either a fundamental solution or a parametrix for the operator of interest. In the discussion which follows we shall treat the calculation of bounds for boundary value problems associated with the Laplace, biharmonic and heat operators. These are operators for which fundamental solutions are known. For more general operators for which a parametrix must be used the technique is similar although in some instances the details become a bit more involved. Such details are fully given in [38], [39], [42]. For another approach, suitable for elliptic operators, see [4].

SECOND ORDER ELLIPTIC PROBLEMS

Consider the fundamental solution for the Laplace operator

$$\Gamma(x,\xi) = \begin{cases} -\frac{1}{2\pi}\log r &, \quad N = 2 \\ \frac{r^{-(N-2)}}{(N-2)\omega_N} &, \quad N \geq 3 \end{cases}$$

where $x = (x_1,\cdots,x_N)$, $\xi = (\xi_1,\dots,\xi_N)$, $r^2 = \sum_{i=1}^{N} (x_i-\xi_i)^2$ and ω_N denotes the surface area of the N-dimensional unit sphere. We assume that the point ξ

is interior to B and that $S_p \subset S_{p'} \subset B$ are spheres of radius p and p', (p < p'), respectively, centered at ξ.

Introduce a C^2 function $\eta = \eta(x)$ defined as follows:

$$\eta(x) = \begin{cases} 1 & , \quad x \in \bar{S}_{p_o} \\ 0 \le \eta \le 1 & , \quad x \in S_{p'} - \bar{S}_{p_o} \\ 0 & , \quad x \in B - S_{p'} \end{cases} \tag{8.1}$$

Initially $p = p_o$ but later we shall take limits as $p \to 0$ and we want to emphasize that the sphere S_{p_o} remains unchanged during this process. Now by Green's second identity we have

$$\int_{B-S_p} [u\Delta(\eta\Gamma) - \eta\Gamma\Delta u]dx = \oint_{\partial B+S_p} [u\partial(\eta\Gamma)/\partial n - \eta\Gamma\,\partial u/\partial n]dS$$

or

$$\int_{S_{p'}-S_p} (uH-\eta\Gamma\Delta u)dx = \oint_{\partial S_p} (u\partial\Gamma/\partial n-\Gamma\partial u/\partial n)dS \tag{8.2}$$

where we have used (8.1). In this equation $H = \Gamma\Delta\eta + 2\Gamma_{,i}\eta_{,i}$ since $\Delta\Gamma = 0$ in $S_{p'} - S_p$. Now on $S_{p'}$, $\partial/\partial n = \partial/\partial r$, so that performing the derivatives indicated in the right hand side of (8.2) and taking the limit as $p \to 0$ it is easily seen that

$$\oint_{S_p} u\frac{\partial\Gamma}{\partial n}\,dS \to u(\xi) \quad , \quad \oint_{\partial S_p} \Gamma\frac{\partial u}{\partial n}\,dS \to 0$$

and thus we have

$$u(\xi) = \int_{S_{p'}-S_p} (uH-\eta\Gamma\Delta u)dx \tag{8.3}$$

Squaring (8.3) and using the Schwarz and a-g inequalities we obtain

$$
\begin{aligned}
|u(\xi)|^2 &\le K_1(\xi) \int_{S_{p'}} u^2 dx + K_2(\xi) \int_{S_{p'}} (\Delta u)^2 dx \\
&\le K_1(\xi) \int_B u^2 dx + K_2(\xi) \int_B (\Delta u)^2 dx
\end{aligned} \tag{8.4}
$$

where

$$
K_1(\xi) = 2 \int_{S_{p'}} H^2 dx \ , \qquad K_2(\xi) = 2 \int_{S_{p'}} (\eta\Gamma)^2 dx \ . \tag{8.5}
$$

The bound (8.4) is useful only for N = 2,3 because it is only for these values that K_2 is finite. For $n > 3$ the fundamental solution is not square integrable and a weighted Schwarz inequality must be used in the last term of (8.4) giving the pointwise bound

$$
|u(\xi)|^2 \le K_1(\xi) \int_B u^2 dx + K_2^{(\alpha)}(\xi) \int_B r^{-\alpha} (\Delta u)^2 dx \tag{8.6}
$$

with K_1 as before and

$$
K_2^{(\alpha)}(\xi) = 2 \int_{S_{p'}} r^{\alpha} (\eta\Gamma)^2 dx \ . \tag{8.7}
$$

Of course, α is chosen so that $K_2^{(\alpha)}$ is finite and $r^{-\alpha}(\Delta u)^2$ is integrable. Using (8.4) or (8.6) a pointwise bound for the difference between an approximation u_* and the solution w of, for instance, the Dirichlet problem

$$
\Delta w = f \quad \text{in} \quad B \ ,
$$

$$
w = g \quad \text{on} \quad \partial B \ ,
$$

is obtained using the inequality of Theorem 4.1 with $u = w - u_*$:

$$
\begin{aligned}
|w(\xi) - u_*(\xi)|^2 &\le K_1(\xi) \left\{ \alpha_1 \int_B (f - \Delta u_*)^2 dx + \alpha_2 \oint_{\partial B} (g - u_*)^2 dS \right\} \\
&\quad + K_2(\xi) \int_B (f - \Delta u_*)^2 dx \ .
\end{aligned} \tag{8.8}
$$

FOURTH ORDER ELLIPTIC PROBLEMS

Drawing on the results of the previous section we quickly sketch the approach to this case. For the biharmonic operator Δ^2 the fundamental solution is [16]:

$$\Gamma(x,\xi) = \frac{1}{2^3\omega_N (\frac{N}{2}-1)(\frac{N}{2}-2)} r^{4-N}$$

for all odd N and for even N > 4, and

$$\Gamma(x,\xi) = \frac{(-1)^{(N-2)/2}}{2^3\pi^{N/2} (2-\frac{N}{2})!} r^{N-4} \log r$$

for even $N \leq 4$. See [16] and [22] for information on parametrices for more general fourth order equations. The fundamental solution has the property that

$$u(\xi) = \lim_{p\to 0} \oint_{\partial S_p} u \frac{\partial(\Delta\Gamma)}{\partial n} dS .$$

Thus, Green's second identity yields

$$u(\xi) = \int_{S_{p'}-S_p} (uH-\eta\Gamma\Delta^2 u)dx$$

where $H = \Gamma\Delta^2\eta + 4\Gamma,_j\Delta\eta,_j + 2\Delta\eta\Delta\Gamma + 4\eta,_{ij}\Gamma,_{ij} + 4\eta,_i\Delta\Gamma,_i$ so that, as above,

$$|u(\xi)|^2 \leq K_1(\xi) \int_B u^2 dx + K_2(\xi) \int_B (\Delta^2 u)^2 dx . \qquad (8.9)$$

Having (8.9) it is now clear how to obtain pointwise bounds in biharmonic boundary value problems for which an appropriate a priori inequality is known.

SECOND ORDER PARABOLIC PROBLEMS

To obtain pointwise bounds in problems involving the heat operator $L = \Delta - \partial/\partial t$, we need the fundamental solution Γ^* of the formal adjoint $L^* = \Delta + \partial/\partial t$:

$$\Gamma^*(x,t;\xi,\tau) = \frac{1}{(2\sqrt{\pi})^N\ (\tau-t)^{N/2}} \exp\left\{ -\frac{\sum(\xi_i-x_i)^2}{4(\tau-t)} \right\}, \qquad \tau > t\ .$$

The function Γ^* has the following properties [15]:

(a) $L^*\Gamma^*(x,t;\xi,\tau) = 0$ for each fixed (ξ,τ) ,

(b) $\lim_{t\to\tau} \int_{B_\tau} \Gamma^*(x,t;\xi,\tau)\ f(\xi)d\xi = f(x)$,

for every continuous function f in B_t .

Now define two subregions D_1 , D_2, of D as follows:

$$D_1 = \{(x,t)\,|\,r = (\sum(x_i-\xi_i)^2 + (\tau-t)^2)^{\frac{1}{2}} \le r_1\ , \qquad t < \tau\}\ ,$$

$$D_2 = \{(x,t)\,|\,r \le r_2, \qquad t < \tau\}\ ,$$

where $r_1 < r_2$ and r_2 is such that D_2 is entirely in D.

Introduce a C^2 function $\eta = \eta(x,t)$ as follows:

$$\eta(x,t) = \begin{cases} 1 & , \quad (x,t) \in \bar{D}_1 \\ 0 \le \eta \le 1 & , \quad (x,t) \in D_2-\bar{D}_1 \\ 0 & , \quad (x,t) \in D-D_2 \end{cases}$$

Then for any C^2 function $u = u(x,t)$, we have, for any $P = (\xi,\tau) \in D_1$,

$$u(P) = \int_D [uL^*(\eta\Gamma^*) - \eta\Gamma L^*u]dV$$

where we have used Green's second identity for L:

$$vLu - uL^*v = (vu,_i),_i - (uv,_i),_i - \partial(uv)/\partial t\ .$$

Setting

$$L^*(\eta\Gamma^*) = \Gamma^*L^*\eta + 2\,\Gamma^*,_i\eta,_i \equiv H(x,t,\xi,\tau)$$

we obtain

$$u(P) = \int_D [uH-\eta\Gamma^* Lu]dV \ .$$

An application of the Schwarz and a-g inequalities then gives the bound

$$|u(P)|^2 \le K_1(P) \int_D u^2 dV + K_2^{(\alpha)}(P) \int_D r^{-\alpha}(Lu)^2 dV$$

where

$$K_1(P) = 2 \int_D H^2 dV \quad \text{and} \quad K_2(P) = 2 \int_D r^{\alpha}(\eta\Gamma^*)^2 dV \ .$$

For problems in only one space dimension we can take $\alpha = 0$ since Γ^* is square-integrable in that case. See [39] for results on the selection of α so that $K_2^{(\alpha)}$ is finite.

9 Applications to eigenvalue estimation

In this chapter we present a method which gives improvable upper and lower bounds for eigenvalues of self-adjoint elliptic operators given only rough preliminary estimates for them. The method applies to the classical membrane and plate eigenvalue problems as well as to Steklov eigenvalue problems. It uses some of the explicit a priori inequalities given in previous chapters, and, as is usually the case when applying methods based on a priori inequalities, the trial functions used in the procedure need not satisfy either the eigenvalue equation or boundary conditions.

Our approach here deviates somewhat from that of previous chapters in that attention will be given to particular problems only after the method has been first developed in a general setting. Further, in order to make the chapter self-contained, the illustrative numerical examples are included along with the theory of the method.

AN A POSTERIORI INEQUALITY

Central to the method is the a posteriori inequality given in Theorem 9.1 below. This inequality is related to one given by Moler and Payne [24] (see also [14], [25]) but differs in the important aspect that it can be combined with a priori inequalities to permit the estimation of eigenvalues in terms of quadratic functionals of "arbitrary" test functions. We shall refer to this combined method as the method of a posteriori-a priori inequalities.

We proceed now to develop the a posteriori inequality. Let A be an operator with domain $\mathfrak{D}(A)$ which is dense in the separable Hilbert space H. Let

A be symmetric, so that

$$(u,Av) = (Au,v) , \qquad u,v \in \mathfrak{D}(A)$$

and let A have pure point spectrum $\{\lambda_i\}$ with corresponding orthonormal eigenvectors $\{u_i\}$ which are complete in H. Let A_* be an extension of A, so that $\mathfrak{D}(A) \subset \mathfrak{D}(A_*) \subset H$ with $A_* u = Au$ for $u \in \mathfrak{D}(A)$.

The a posteriori inequality is given in the following theorem.

Theorem 9.1. Assume the above hypotheses. For any number λ_*, and any $u_* \in \mathfrak{D}(A_*)$, suppose there exists a function w satisfying

$$A_* w = A_* u_* - \lambda_* u_* \tag{9.1}$$

Then

$$\min_i \left| \frac{\lambda_i - \lambda_*}{\lambda_i} \right| \le \frac{\|w\|}{\|u_*\|} \equiv \epsilon \tag{9.2}$$

and thus if $\epsilon < 1$ there exists an eigenvalue λ_k of A satisfying

$$\frac{|\lambda_*|}{1+\epsilon} \le |\lambda_k| \le \frac{|\lambda_*|}{1-\epsilon} \quad . \tag{9.3}$$

Proof: By symmetry and (9.1)

$$\lambda_i (u_* - w, u_i) = (u_* - w, Au_i) = (Au_* - Aw, u_i) = \lambda_* (u_*, u_i)$$

so that

$$(\lambda_i - \lambda_*)(u_*, u_i) = \lambda_i (w, u_i) ,$$

or

$$\frac{\lambda_i - \lambda_*}{\lambda_i} = \frac{(w, u_i)}{(u_*, u_i)} \quad .$$

Now since $\{\lambda_i\}$ has no finite limit point, there exists some k such that

$$|\lambda_k - \lambda_*| / |\lambda_k| = \min_i |\lambda_i - \lambda_*| / |\lambda_i|$$

and consequently, for this k,

$$\frac{|\lambda_k - \lambda_*|}{|\lambda_k|} |(u_*, u_i)| \le |(w, u_i)|$$

for all i, and thus

$$\frac{|\lambda_k - \lambda_*|^2}{|\lambda_k|^2} \sum_{i=1}^{\infty} |(u_*, u_i)|^2 \le \sum_{i=1}^{\infty} |(w, u_i)|^2 \quad . \tag{9.4}$$

We now use the completeness of the $\{u_i\}$ which implies that $\|w\|^2 = \sum_i |(w, u_i)|^2$, $\|w\|^2 = \sum_i |(u_*, u_i)|^2 = \|u_*\|^2$, in (9.4) to obtain

$$\frac{|\lambda_k - \lambda_*|^2}{|\lambda_k|^2} \le \frac{\|w\|^2}{\|u_*\|^2} \tag{9.5}$$

thus proving the theorem.

It is not desirable to have to actually obtain the function w which appears in the theorem and it is at this point that we introduce an appropriate a priori inequality which estimates w in terms of u_*.

AN EXAMPLE — THE FIXED MEMBRANE PROBLEM

Let B be a bounded region of Euclidean N-space with boundary ∂B. Let the Hilbert space H be $\mathfrak{L}_2(B)$, the space of functions which are square integrable on B. Let A be the negative Laplacian $-\Delta$ with domain $\mathfrak{D}(A)$ the twice differentiable functions vanishing on ∂B. Thus A_* will be $-\Delta$ with domain $\mathfrak{D}(A_*)$ the functions which are just twice differentiable on B.

This gives the classical fixed membrane problem

$$-\Delta u = \lambda u \quad \text{on} \quad B, \qquad u = 0 \quad \text{on} \quad \partial B \ , \tag{9.6}$$

whose eigenvalues and eigenvectors satisfy the hypotheses of Theorem 9.1. The appropriate *a priori* inequality to use is that given on page 22

$$\|w\|^2 = \int_B w^2\,dx \le 2\lambda_1^{-2} \int_B (\Delta w)^2\,dx + 2q_1^{-1} \oint_{\partial B} w^2 dS \tag{9.7}$$

for $w \in \mathfrak{D}(A_*)$. Combining this with (9.5) gives

$$\frac{|\lambda_k - \lambda_*|^2}{|\lambda_k|^2} \le \frac{2\lambda_1^{-2} \int_B (\Delta u_* + \lambda u_*)^2\,dx + 2q_1^{-1} \oint_{\partial B} u_*^2 dS}{\int_B u_*^2 dx} \ . \tag{9.8}$$

Now the right side of (9.8) is a ratio of quadratic forms in the arbitrary twice-differentiable function u_*. Thus we can let u_* be a linear combination of test functions, say

$$u_* = \sum_{\ell=1}^{n} a_\ell \varphi_\ell \tag{9.9}$$

and minimize the right side of (9.8) with respect to the coefficients a_k as in the Rayleigh-Ritz method. This leads to the relative matrix eigenvalue problem

$$Ma - \epsilon^2\, Na = 0 \ , \tag{9.10}$$

where

$$M = [2\lambda_1^{-2} \int_B (\Delta\varphi_i + \lambda_* \varphi_i)(\Delta\varphi_j + \lambda_* \varphi_j)\,dx + 2q_1^{-1} \oint_{\partial B} \varphi_i \varphi_j dS] \ ,$$

$$N = [\int_B \varphi_i \varphi_j dx] \ ,$$

and

$$a = (a_1, \cdots, a_n)^T \ .$$

Now let ϵ be the smallest eigenvalue of (9.10), then

$$\left| \frac{\lambda_k - \lambda_*}{\lambda_k} \right| \le \epsilon$$

or, if $\epsilon < 1$,

$$\lambda_*/(1+\epsilon) \le \lambda_k \le \lambda_*/(1-\epsilon) \ , \tag{9.11}$$

giving upper and lower bounds for the fixed membrane eigenvalue λ_k which is closest to λ_*.

OTHER EIGENVALUE PROBLEMS

The free membrane eigenvalue problem is

$$\Delta u + \mu u = 0 \quad \text{in} \quad B \ ,$$

$$\frac{\partial u}{\partial n} = 0 \quad \text{on} \quad \partial B \ ,$$

with eigenvalues $0 = \mu_1 < \mu_2 \le \mu_3 \le \cdots$. Combining inequality (9.2) with the a priori inequality of Theorem 4.2 yields

$$\min_{i \ne 1} \left| \frac{\mu_* - \mu_i}{\mu_i} \right|^2 \le \frac{\alpha_1 \int_B (\Delta u_* + \mu_* u_*)^2 dx + \alpha_2 \oint_{\partial B} \left(\frac{\partial u_*}{\partial n}\right)^2 dS}{\int_B u_*^2 dx} \tag{9.12}$$

for any number μ_* and any function u_* satisfying

$$\int_B u_* dx = 0 \ .$$

The clamped plate eignevalue problem is

$$\Delta^2 u - \Omega u = 0 \quad \text{in} \quad B \ ,$$

$$u = \frac{\partial u}{\partial n} = 0 \quad \text{on} \quad \partial B \ ,$$

with eigenvalues $0 < \Omega_1 \le \Omega_2 \le \cdots$. Combining (9.2) with the a priori inequality of Theorem 7.1 yields

$$\min_i \left|\frac{\Omega_*-\Omega_i}{\Omega_i}\right|^2 \int_B u^2\,dx$$

$$\le \alpha_1 \int_B (\Delta^2 u_* - \Omega_* u_*)^2\,dx + \alpha_2 \oint_{\partial B} u_*^2\,dS + \alpha_3 \oint_{\partial B} \left(\frac{\partial u_*}{\partial n}\right)^2 dS + \qquad (9.13)$$

$$+ \alpha_4 \oint_{\partial B} \left(\frac{\partial u_*}{\partial s}\right)^2 dS\ ,$$

for any number Ω_* and any function u_*. Here $\partial/\partial s$ is the tangential derivative on ∂B.

A NUMERICAL EXAMPLE

We now discuss the results obtained by applying the method to rhombical membranes. This region was selected for two reasons. One is that very good upper and lower bounds for selected eigenvalues have already been given by Stadter [47] thus giving us a means of comparing our results. Another reason is that as the region becomes more elongated in one direction good bounds are more difficult to obtain. Thus for small values of θ (θ denotes the least interior angle, see Fig. 1), all methods start giving poor bounds and we wanted to see how well our method would do in this relatively difficult situation.

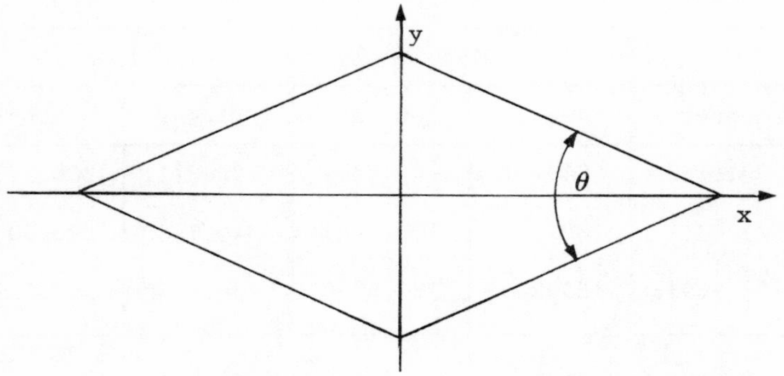

FIGURE 1

Stadter used the well-regarded method of special choice [13] with carefully constructed trial functions which satisfy the boundary conditions of the problem, i.e., they vanish on the boundary. On the other hand, desiring to take advantage of the generality in the selection of the trial functions which our method allows, we chose as trial functions simple products of even powers of x and y giving

$$u_* = \sum_{m=0}^{M} \sum_{n=0}^{N} a_{mn} x^{2m} y^{2n} \tag{9.13}$$

or products of cos $m\pi x$ and cos $n\pi y$ giving

$$u_* = \sum_{m=0}^{M} \sum_{n=0}^{N} b_{mn} \cos m\pi x \cos n\pi y \; . \tag{9.14}$$

Such choices restrict us to approximating eigenvalues of the even-even symmetry class but it is this class which contains the lowest, and usually the most physically important, eigenvalue λ_1.

The results of our computations are given in Table 1 and for comparison those of Stadter are given in Table 2. For $\theta = 75°$ and $45°$ the test function (9.13) was used with $M = N = 8$, while for $\theta = 15°$ the function (9.14) was used with $M = N = 6$.

θ	75°		45°		15°	
i	lower	upper	lower	upper	lower	upper
1	20.8660	20.8782	34.7017	34.8569	196.8111	206.7539
2	79.0234	79.0711	100.1116	100.4642	340.7600	386.9026
3	108.8175	108.9697	183.3045	187.6383	436.9660	652.2823

TABLE 1. Bounds for eigenvalues of unit side rhombical membranes in even-even symmetry class.

θ	75°		45°		15°	
i	lower	upper	lower	upper	lower	upper
1	20.8613	20.8871	34.7113	35.0283	199.0205	207.6664
2	78.9173	79.2392	100.1073	101.2326	357.7928	390.1554
3	108.7433	109.2763	184.1569	189.7827	371.5017	708.1836

TABLE 2. Stadter's bounds for eigenvalues of unit side rhombical membrane in even-even symmetry class.

10 Numerical examples

We now present some numerical examples which illustrate the usefulness of explicit *a priori* inequalities in computing approximate solutions for boundary value problems.

We have mentioned in the introduction that the user must supply a set of trial functions from which the approximating solution is constructed. However, these functions *need not satisfy either the differential equation or the boundary conditions*. This can be a real advantage in those problems for which functions which satisfy the boundary conditions are difficult to construct and in those problems for which solutions of the differential equation are not known. This freedom in selection of the trial functions presents the opportunity to incorporate known information about the problem into the approximate solution. For instance, one could use trial functions suggested by the boundary conditions, the differential equation or solutions of special cases of the problem which can be solved exactly. See [12, p.34] for further discussion along these lines. Generally, if functions which satisfy either the differential equation or boundary conditions are known they should be used for then some terms in the inequality drop out. However, in order to illustrate that good approximations are possible with trial functions which do not satisfy any of the prescribed data, we shall use such trial functions in most of our examples.

AN ELLIPTIC EXAMPLE

As a first example consider the problem

$$\Delta w = -1 \quad \text{in} \quad B, \qquad w = 0 \quad \text{on} \quad \partial B ,$$

where B is the rectangular region: $\{(y,y) | -a < x < a, -b < y < b\}$ (In this and all other examples we use x and y instead of x_1 and x_2.) This problem has as one physical interpretation the description of the velocity for fully developed flow of a Newtonian viscous fluid in a straight rectangular channel of aspect ratio a/b. An exact solution of this problem has been given by O'Brien [26] as

$$\frac{w(x,y)}{w_1} = 2 - \frac{x^2}{a^2} - \frac{y^2}{b^2} - \sum_{n=0}^{\infty} 4(-1)^n \left\{\frac{\cosh p_n x}{\cosh p_n a} \cos p_n y + \frac{\cosh q_n y}{\cosh q_n b} \cos q_n \mathbf{x}\right\} . \tag{10.1}$$

where

$$w_1(a,b) = ab/(a^2+b^2), \ \sigma_n = 2/[(2n+1)\pi], \ p_n = (\sigma_n b)^{-1}, \ q_n = (\sigma_n a)^{-1} .$$

As trial functions we used products of even powers of x and y. Thus our approximating function u_a is given by

$$u_a(x,y) = \sum_{m=1}^{N_x} \sum_{n=1}^{N_y} a_{mn} x^{2m-2} y^{2n-2} . \tag{10.2}$$

The appropriate inequality to use is that given on page 22, i.e.,

$$\int_B u^2 dx \le 2q_1^{-1} \oint_{\partial B} u^2 dS + 2\lambda_1^{-2} \int_B (\Delta u)^2 dx .$$

For the rectangular region considered $\lambda_1 = \pi^2/4 \ (1/a^2+1/b^2)$. A lower bound for q_1 has been given by Kuttler (Remarks on a Stekloff eigenvalue problem, SIAM J. Numer. Anal. 9 (1972), 1-5), for a square of side a by $q_1 \ge 2.96 \ a^{-1}$. Results of calculations for $a = b = 1$, $N_x = N_y = 5$ are given in Table 1 where we have also included the actual solution computed using the first forty terms of (10.1). Agreement between the actual and approximate solutions are quite good, at least once we are slightly removed from the boundary. It is worth noting that even the so-called exact

TABLE 1

Values of the approximating solution u_a and the actual solution u at various values of y for $x = 0, \frac{1}{2}$.

y	$u_a(0,y)$	$u(0,y)$	$u_a(\frac{1}{2},y)$	$u(\frac{1}{2},y)$
1.00	-0.295023(-4)	-0.173537(-5)	-0.101495(-3)	0.172371(-6)
0.96	0.262113(-1)	0.262214(-1)	0.216465(-1)	0.217245(-1)
0.92	0.508995(-1)	0.508963(-1)	0.418722(-1)	0.419317(-1)
0.88	0.740805(-1)	0.740775(-1)	0.606588(-1)	0.607033(-1)
0.84	0.958347(-1)	0.958176(-1)	0.780869(-1)	0.781189(-1)
0.80	0.116188	0.116168	0.942332(-1)	0.942550(-1)
0.76	0.135200	0.135180	0.109171	0.109185
0.72	0.152921	0.152901	0.122968	0.122977
0.68	0.169399	0.169381	0.135690	0.135696
0.64	0.184680	0.184663	0.147397	0.147401
0.60	0.198808	0.198793	0.158143	0.158148
0.56	0.211824	0.211811	0.167981	0.167988
0.52	0.223768	0.223757	0.176957	0.176966
0.48	0.234678	0.234668	0.185113	0.185124
0.44	0.244586	0.244578	0.192488	0.192499
0.40	0.253526	0.253519	0.199116	0.199128
0.36	0.261525	0.261520	0.205027	0.205039
0.32	0.268612	0.268607	0.210248	0.210258
0.28	0.274808	0.274805	0.214803	0.214811
0.24	0.280136	0.280134	0.218711	0.218716
0.20	0.284614	0.284612	0.221991	0.221992
0.16	0.288256	0.288256	0.224655	0.224652
0.12	0.291076	0.291076	0.226716	0.226710
0.08	0.293083	0.293084	0.227181	0.228173
0.04	0.294284	0.294285	0.229058	0.229048
0.00	0.294684	0.294685	0.229350	0.229340

solution is, in an important (numerical) sense, approximate since in its evaluation a limiting process in involved. Thus, even though it should evaluate to zero on the boundary what we obtain in actual calculation is something other than zero. In Table 2 we give results for $N_x = N_y = 7$ and $N_x = N_y = 9$ for certain values of y and for $x = \frac{1}{2}$. For those values of y which appear in Table 1 but not in Table 2, the approximate and actual solutions agreed to at least six significant figures and thus are not given. Table 2 illustrates the obvious, i.e., the approximation improves, particularly near the boundary, as more terms are used. Table 3 gives the coefficients a_{mn} of approximate solution u_a. Also we have the following norm error bounds:

For $N_x = N_y = 5$

$$\int_B (u-u_a)^2 \, dxdy \leq 1.0616(-6)^{\dagger} \quad ,$$

for $N_x = N_y = 7$

$$\int_B (u-u_a)^2 \, dxdy \leq 6.1613(-8) \quad ,$$

for $N_x = N_y = 9$

$$\int_B (u-u_a)^2 \, dxdy \leq 5.9819(-9) \quad .$$

A PARABOLIC EXAMPLE

To illustrate the use of the inequalities which have applications to parabolic initial-boundary value problems we consider the problem

$$\frac{\partial^2 w}{\partial x^2} - \frac{\partial w}{\partial t} = 0 \quad \text{in } D = \{x \mid -\pi<x<\pi\} \times \{t \mid 0<t< T< \infty\} \ , \tag{10.3}$$

---

†In this chapter we use the notation that $(-6) \equiv \times 10^{-6}$

y	$N_x = N_y = 7$	$N_x = N_y = 9$
1.00	-0.554427 (-4)	-0.710377 (-5)
0.96	0.217586 (-1)	0.217208 (-1)
0.92	0.419527 (-1)	0.419300 (-1)
0.88	0.607161 (-1)	0.607025 (-1)
0.84	0.781266 (-1)	0.781185 (-1)
0.80	0.942596 (-1)	0.942549 (-1)
0.76	0.109187	
0.72	0.122978	

TABLE 2. Values of u_a near the boundary for $x = \frac{1}{2}$

$x^p y^q$	a_{pq}
0,0	0.294684
(0,2),(2,0)	-0.249937
(0,4),(4,0)	-0.457160 (-1)
(0,6),(6,0)	0.328658 (-3)
(0,8),(8,0)	0.610867 (-3)
(2,2)	0.271657
(2,4),(4,2)	0.116332 (-2)
(2,6),(6,2)	-0.313065 (-1)
(2,8),(8,2)	0.112125 (-1)
(4,4)	0.119341
(4,6),(6,4)	-0.102545
(4,8),(8,4)	0.509579 (-2)
(6,6)	0.289340
(6,8),(8,6)	-0.106834
(8,8)	0.596247 (-1)

TABLE 3. Coefficients of the approximating solution u_a for $N_x = N_y = 5$.

$$w(x,0) = \cos x \quad \text{on} \quad B = \{x \mid -\pi < x < \pi\} , \tag{10.4}$$

$$w(\pm\pi,t) = e^{-t} \quad \text{on} \quad S = \{(x,t) \mid x = \pm\pi, 0 < t \le T\} . \tag{10.5}$$

As trial functions we use the polynomials

$$\varphi_n(x,t) = \sum_{k=1}^{n} \frac{x^{2k-2}\, t^{n-k}}{(2k-2)!\,(n-k)!} , \qquad n=1,2,\cdots ,$$

each of which satisfies (10.3), and thus the approximation function

$$u_a = \sum_{n=1}^{N} a_n \varphi_n(x,t) , \tag{10.6}$$

where the a_n's are to be determined, also satisfies (10.3). The approximate inequality to use for the problem is that of Theorem 5.1:

$$\int_D u^2\,dxdt \le \alpha_1 \int_D (Lu)^2\,dxdt + \alpha_2 \int_B u^2\,dx + \alpha_3 \int_S u^2\,d\sigma \tag{10.7}$$

where $L = \dfrac{\partial^2}{\partial x^2} - \dfrac{\partial}{\partial t}$. The constants α_1, α_2, α_3 are given by

$$\alpha_1 = 2\lambda_1^{-2}$$

$$\alpha_2 = 2\lambda_1^{-1}$$

$$\alpha_3 = p_m^{-1} \bar{K} ,$$

where p_m and $\bar{K}$ are as defined on page 34.

Since (10.6) satisfies (10.3), the first term on the right hand side of (10.7) drops out. Thus we need only compute the explicit constants α_2 and α_3.

The value for α_2 is easily obtained since for B, $\lambda_1 = 1$ and thus $\alpha_2 = 2$. The expression for α_3 also reduces to a simple expression if we take $f^1 \equiv x$, and $f^2 \equiv f^t = 1$. Then in $\bar{K}$, p. 34 we have

$$C = 1 ,$$

$$|f^{N+1}|_m \equiv |f^t| = 1 ,$$

$$|f^i f^i|_M^{\frac{1}{2}} = |x^2+1|_M^{\frac{1}{2}} = (\pi^2+1)^{\frac{1}{2}} ,$$

$$\left|\frac{f^{N+1}_{,i}\, f^{N+1}_{,i}}{f^{N+1}}\right|_M = 0 ,$$

$$\left|\frac{f^i f^i}{f^{N+1}}\right|_M = |x^2+1|_M = \pi^2 + 1 ,$$

so that

$$\bar{K} = 3 + (\pi^2+1)^{\frac{1}{2}} (1+2(\pi^2+1)^{\frac{1}{2}})$$

and then

$$\alpha_3 = 6 + 2(\pi^2+1)^{\frac{1}{2}} (1+2(\pi^2+1)^{\frac{1}{2}}) .$$

Finally, setting $u = w - u_a$, where w denotes the solution of (10.3)-(10.5), we have that

$$\int_{-\pi}^{\pi}\int_0^T (w- \sum_{n=1}^N a_n\varphi_n)^2 dxdt \le 2 \int (\cos x - \sum_{n=1}^N a_n\varphi_n(x,0))^2 dx$$
$$+ \{6+2(\pi^2+1)^{\frac{1}{2}} (1+2(\pi^2+1)^{\frac{1}{2}})\}\int_S (e^{-t} - \sum_{n=1}^N a_n\varphi_n(\pm\pi,t))^2 dt .$$

This expression is now minimized with respect to the a_n, as explained in the introduction.

The results of our calculations are given in Table 4 where we have also included pointwise bound calculations as discussed in Chapter 8, and for comparison, the values of the exact solution given by $w(x,t) = e^{-t}\cos x$. Several observations concerning these calculations are of interest:

x	t	Approximate Solution	Error Bounds on Approxi.	Exact Solution $u = e^{-t} \cos x$
0	0.2	0.8187307	0.0000583	0.8187307
.2π	0.2	0.6623671	0.0000583	0.6623671
.4π	0.2	0.2530017	0.0000583	0.2530017
.6π	0.2	-0.2530017	0.0000583	-0.2530017
.8π	0.2	-0.6623671	0.0000583	-0.6623671
0	0.6	0.5488116	0.0000110	0.5488116
.2π	0.6	0.4439980	0.0000110	0.4439980
.4π	0.6	0.1695921	0.0000110	0.1695921
.6π	0.6	-0.1695921	0.0000110	-0.1695921
.8π	0.6	-0.4439980	0.0000110	-0.4439980
0	1.0	0.3678794	0.0000072	0.3678794
.2π	1.0	0.2976207	0.0000072	0.2976207
.4π	1.0	0.1136810	0.0000072	0.1136810
.6π	1.0	-0.1136810	0.0000072	-0.1136810
.8π	1.0	-0.2976207	0.0000142	-0.2976207
0	1.5	0.2231302	0.0000123	0.2231302
.2π	1.5	0.1805161	0.0000123	0.1805161
.4π	1.5	0.6895101(-1)	0.0000123	0.6895101(-1)
.6π	1.5	-0.6895100(-1)	0.0000154	-0.6895101(-1)
.8π	1.5	-0.1805160	0.0000407	-0.1805161
0	2.0	0.1353353	0.0000127	0.1353353
.2π	2.0	0.1094885	0.0000127	0.1094885
.4π	2.0	0.4182087(-1)	0.0000145	0.4182090(-1)
.6π	2.0	-0.4182100(-1)	0.0000228	-0.4182090(-1)
.8π	2.0	-0.1094886	0.0000620	-0.1094885
0	2.5	0.8208500(-1)	0.0000253	0.8208500(-1)
.2π	2.5	0.6640812(-1)	0.0000253	0.6640816(-1)
.4π	2.5	0.2536559(-1)	0.0000374	0.2536566(-1)
.6π	2.5	-0.2536571(-1)	0.0000590	-0.2536566(-1)
.8π	2.5	-0.6640818(-1)	0.0001556	-0.6640816(-1)
0	3.0	0.4978704(-1)	0.0000758	0.4978707(-1)
.2π	3.0	0.4027856(-1)	0.0000934	0.4027859(-1)
.4π	3.0	0.1538495(-1)	0.0001381	0.1538505(-1)
.6π	3.0	-0.1538505(-1)	0.0002178	-0.1538505(-1)
.8π	3.0	-0.4027803(-1)	0.0005743	-0.4027859(-1)
0	4.0	0.1831575(-1)	0.0004035	0.1831564(-1)
.2π	4.0	0.1481759(-1)	0.0004972	0.1481766(-1)
.4π	4.0	0.5659724(-2)	0.0007355	0.5659844(-2)
.6π	4.0	-0.5659134(-2)	0.0011600	-0.5659844(-2)
.8π	4.0	-0.1481537(-1)	0.0030588	-0.1481766(-1)

TABLE 4*

*Approximate values for negative x are obtained by reflecting those given across the t-axis since the solution is symmetric with respect to the t-axis.

(A) The approximate values are very close to the actual values; in many cases approximate and actual values agree to seven significant figures. Agreement between approximate and actual values is always much better than the error bounds indicate. Thus although the error bounds are generally good when looked at as a percentage of the approximating value, they are pessimistic when compared to actual errors.

(B) As is evident from the discussion in Chapter 8, the pointwise error bounds worsen as the boundary is approached ($K_1(P)$ and $K_2(P)$ become unbounded as P approaches the boundary). This is noticeable at the larger values of t. Notice, however, that the approximations themselves are not so severely affected.

(C) For $t \geq 3.0$ the error bounds, and to a much smaller degree the approximations, become progressively worse. This happens because as time increases more trial functions are needed to construct the approximating function if a given error is to be maintained. In our calculations we restricted N to be less than or equal to 10.

With the exception of the values obtained for $t = 0.2$ and 0.4, all results in Table 4 were obtained using ten trial functions. At $t = 0.2$, eight functions were employed while nine were used for $t = 0.4$. This was necessary because the system formed in applying the minimization procedure tended to be slightly ill-conditioned, especially for the smaller values of t. Thus for the first two values of t, calculations employing ten trial functions yielded worse results than calculations employing fewer functions. This tendency toward ill-conditioned systems persisted throughout the calculations although it was not as pronounced at the larger t values. Because of this tendency, and also to avoid loss of significance due to the subtraction of two nearly equal numbers when mean square errors on the boundary were computed, all calculations were done using double precision arithmetic. It is extremely important that highly accurate methods are used to determine the a_n's since the error bounds, but not the approximations, are quite sensitive to errors in these constants.

One last obvious observation concerns the selectivity of the method. That is, an approximate value can be calculated at a few points without the need to perform calculations at many additional points in which one has no interest. Thus if one is interested in having an approximation at the point (x_1, t_1), the approximation can be computed immediately without "building up" the solution through a succession of calculations from $t = 0$ to $t = t_1$.

SOME EXPERIMENTS IN TRIAL FUNCTION SELECTION

From the previous examples it is evident that the accuracy of the approximation will be highly dependent upon how closely the set of trial functions can approximate prescribed data in the norm induced by the a priori inequality. The situation is usually this: we have selected our trial functions from a complete set R of functions, i.e., products of sines and cosines, products of powers of x and y, polynomials which satisfy the differential equation, etc. From this set of trial functions some, or even most, may not be useful in approximating the prescribed data. It would be very desirable if once the minimization process has been done we could in some way use the results thus obtained to select those functions from our set of trial functions which are of most use in the approximation, add more terms of that type, discard those which appear to be of no use and rerun the procedure. Some indication of the importance of the trial functions can be derived from an analysis of the relative values of the coefficients of each trial function and their effect on the computed norm error bounds.

We now discuss the results of some experiments along these lines via the treatment of a biharmonic boundary value problem. The inequality used in the computations is that given in Theorem 7.1. Consider the problem

$$\Delta^2 w = 4 \cos x \sin y \tag{10.8}$$

on the unit square with u and $\partial u/\partial n$ chosen on the boundary so that the solution of the boundary value problem is $u(x,y) = \cos x \sin y$. We mention that from the boundary conditions and (10.8) a natural first guess for the trial functions would be the products $\cos \{{x \atop y}\} \sin \{{y \atop x}\}$. These would have given excellent results since the actual solution is included in this set. To better illustrate our ideas, we consider instead trial functions of the form $x^m y^n$ as providing a "general" set of trial functions which lead to a reasonable problem to analyze and program. In addition, this group of trial functions contains many functions which are not in the power series expansion of the solution and hence would not be expected to be useful in an approximate solution. Also, notice that except for the first few terms, none of these functions satisfy either the differential equation or the boundary conditions. With these trial functions the approximating function becomes

$$u_a(x,y) = \sum_{n=1}^{N} \sum_{m=1}^{N} a_{m+N(n-1)} x^{m-1} y^{n-1} . \tag{10.9}$$

Case 1. We chose N = 3 in (10.9), thus obtaining an approximating function with 9 terms. Of these 9 terms, only 2 appear in the power series expansion of u. The minimization procedure yielded the function (all coefficients have been rounded to 2 significant figures)

$$u_a(x,y) = .012 + .0027x + .0012x^2 + 1.0y + .047xy - .53x^2y - .22y^2 - .094xy^2 + .19x^2y^2 .$$

Case 2. This was the same as Case 1 except the first three trial functions 1, x, x^2 were deleted because they did not appear to be important judging by the values of their coefficients. This case resulted in the following approximating function:

$$u_a(x,y) = 1.0y + .051xy - .53x^2y - .22y^2 - .096xy^2 + .19x^2y^2 ,$$

with very little change resulting from the previous case as would be expected if the deleted trial functions were not important. Nothing more would have been gained here except for the fact that we also started computing the norm error terms and found that 98 percent of the total error was due to the error in approximating $\Delta^2 w$, while the error due to the approximation of the three boundary terms contributed the remaining 2 percent. This suggested that we add to the approximating function more of the terms of the expansion of $\Delta^2 w$, i.e., cos x sin y.

Case 3. The six trial functions of Case 2 were used plus x^4y, x^6y, y^3, x^2y^3, x^4y^3, x^6y^3. This resulted in

$$u_a(x,y) = 1.0y + 0.28xy - .53x^2y + .0025y^2 - .031xy^2 + .00033x^2y^2 + .049x^4y + .0044x^6y - .16y^3 + .12x^2y^3 - .025x^4y^3 + .00061x^6y^3 \qquad (10.10)$$

with a decrease in the norm error bound of two orders of magnitude, while the contribution to the norm error due to the approximation of $\Delta^2 w$ dropped to 13 percent of the total error. But just as important is what happened to the coefficients of xy, y^2, xy^2 and x^2y^2 (which do not appear in the power series expansion of the solution) from Case 3 to Case 2, as compared to the coefficients of y and x^2y (which do appear in the series expansion of the solution). Whereas the coefficients of the latter did not change (at least to the two figures reported here), the former ones decreased in absolute value by at least a factor of one-half and some by several orders of magnitude. This suggested that these terms are not important to the approximation and should be deleted. Doing this results in Case 4.

Case 4. This is the previous case with xy, y^2, xy^2 and x^2y^2 deleted. This resulted in

$$u_a(x,y) = 1.0y - .54x^2y + .057x^4y + .0045x^6y$$
$$- .16y^3 + .11x^2y^3 - .025x^4y^3 + .00062x^6y^3 .$$

At this point it is clear what additional trial functions should be added if a better approximation is desired.

Although this example is in some sense a special case, the results of the experiment lend a good deal of support to our hope that the results of the minimization process, via analysis of the relative values of the coefficients of the trial functions and the norm error bounds, can help in selecting the appropriate trial functions from a large number of candidates. We now turn to more detailed numerical examples.

A BIHARMONIC EXAMPLE

As our last example we consider some biharmonic boundary value problems defined on three regions: a square, a triangle and the quadrant of a circle. The first two regions were chosen for their simplicity, the last because it represents a region which presents some difficulty. For each region we chose the boundary values so that the test problem had as its solution $u(x,y) = e^x + \cos y$. As in the previous section the appropriate inequality to use is that given in Theorem 7.1:

$$\int_B u^2\,dxdy \le \alpha_1 \int_B (\Delta^2 u)^2\,dxdy + \alpha_2 \oint_{\partial B} u^2\,dS + \alpha_3 \oint_{\partial B} (\partial u/\partial n)^2\,dS + \alpha_4 \oint_{\partial B} (\partial u/\partial s)^2\,dS \quad .$$

As the approximation function we used

$$u_a(x,y) = a_0 + \sum_{n=1}^{N} a_n x^n + \sum_{n=1}^{N} a_{n+N} y^n \tag{10.12}$$

Again we point out that except for x^n, y^n, i=1,2,3, none of the trial functions satisfy the differential equation or the boundary values, and extraneous terms are present via odd powers of y.

(A) The square. Consider the biharmonic boundary value problem

$$\Delta^2 w = e^x + \cos y \quad \text{in} \quad B = \{(x,y)\,|\,0<x<\pi,\ 0<y<\pi\}\ , \tag{10.13}$$

$$\left.\begin{aligned} w &= e^x + 1 \\ \partial w/\partial n &= 0 \end{aligned}\right\} \text{ on } \partial B_1 = \{(x,0)\,|\,0\le x<\pi\}\ , \tag{10.14}$$

$$\left.\begin{aligned} w &= e^\pi + \cos y \\ \partial w/\partial n &= e^\pi \end{aligned}\right\} \text{ on } \partial B_2 = \{(\pi,y)\,|\,0\le y<\pi\}\ , \tag{10.15}$$

$$\left.\begin{aligned} w &= e^x - 1 \\ \partial w/\partial n &= 1 + \cos y \end{aligned}\right\} \text{ on } \partial B_3 = \{(x,\pi)\,|\,0<x\le\pi\}\ , \tag{10.16}$$

$$\left.\begin{aligned} w &= 1 + \cos y \\ \partial w/\partial n &= -1 \end{aligned}\right\} \text{ on } \partial B_4 = \{(0,y)\,|\,0<y\le\pi\}\ . \tag{10.17}$$

We now must compute the values of α_1, α_2, α_3, α_4. The constant α_1 is easily obtained since for the square B, $\lambda_1 = 2$, and thus (see p.61)

$$\alpha_1 = 0.25\ .$$

Next chose $f^1 = x - \pi/2$, $f^2 = y - \pi/2$. Then $f^i n_i > 0$ on ∂B, C = 0 (p.21),

$$p_m = \pi/2,\ |f^\ell_{,i} f^\ell_{,i}|_M = \sqrt{2},\ |(\Delta f^\ell)(\Delta f^\ell)|_M = 0,\ |f^\ell_{,\ell}|_M = 2,\ |f^\ell f^\ell|^{\frac{1}{2}}_M = \pi/\sqrt{2}$$

so that $K = (2\sqrt{2} + 1 + 2\pi/\sqrt{2})\,\pi^{-1}$ (eq. 7.20, p.60) and thus

$$\alpha_3 = 8(2\sqrt{2} + 1 + 2\pi/\sqrt{2})\,\pi^{-1} .$$

In a similar manner we find that

$$\alpha_2 = 4(1+\pi^2)/\pi ,$$

$$\alpha_4 = 320K .$$

The results of computations for various values of N are given in Table 5.

N	Norm error bounds	Max. rel. err.	Comp. time (sec)	Size of system to be solved
4	991	1.48	0.34	9
5	143	9.03(-2)	0.42	11
6	8.5	9.70(-3)	0.44	13
7	3.00(-1)	2.80(-3)	0.49	15
8	7.25(-3)	8.37(-4)	0.55	17
9	1.17(-4)	8.28(-6)	0.63	19
10	1.49(-6)	6.08(-7)	0.73	21
11	1.29(-8)	1.85(-8)	0.85	23
12	1.00(-9)	1.49(-9)	0.90	25
13	1.65(-9)	1.62(-9)	1.04	27
14	3.21(-8)	1.29(-8)	1.18	29
15	3.72(-6)	1.65(-7)	1.58	31

Table 5[†]

The actual solution (u) and the approximation (u_a) were computed at the points $x = i\pi$, $y = j\pi$, $i,j=0,0.1,0.2,\cdots,1.0$ and then the pointwise relative error $(u-u_a)/u$. The absolute value of the largest relative error for each N is reported in the third column of Table 5 while the computation times

[†] For any point P at least 0.2 units away from the boundary of B a pointwise bound on the solution is given by multiplying the norm error bound by 1×10^2. This pointwise bound is usually pessimistic by several orders of magnitude.

are given in the fourth column. Notice that computation times are very short, generally less than a second, and only 1.58 second for the largest case (we used the IBM 360/91 at the APL Computer Center). From Table 5 it is seen that both the error bound and relative error decrease with increasing N, as expected, until N = 13 where the error bounds begin to grow quite rapidly with increasing N. The reason for this behavior is that the linear system which arises in the minimization process has a tendency to become ill-conditioned with increasing N, and by the time N = 13, the system is so poorly conditioned that the results of the computations rapidly lose significance. This does not appear to be a real deterrent to the use of these methods since excellent results are obtainable from relatively small problems, as Table 5 illustrates, which are not bothered by ill-conditioning.

In Table 6, for the case N = 10, we give the minimizing values of the a_i along with the corresponding coefficients in the power series expansion of of e^x and cos y.

(B) The triangle. The next region considered was the right triangle lying in the first quadrant with legs of length π lying along the positive x- and y-axis. Again the boundary values were chosen so that the resulting boundary value problem had as solution $u(x,y) = e^x + \cos y$, and the approximation functions (10.12) were used. Except for the necessity to evaluate slightly more complicated integrals than for the square, there was nothing new about the calculations here.

The results for our arbitrarily selected "standard case", i.e., N = 10 in (10.12) are:

norm err. bound = 3.19×10^{-7}

max. rel. err. = 1.50×10^{-5}

computation time = 0.64 sec.

The coefficients of the trial functions for this case are the second column of a_i's of Table 6.

n	a_n Square	Triangle	Quadrant	Correspond-ing coeff.in power series expan.of e^x	Correspond-ing coeff. in power series expan.of cos y
0	2.00000	2.00000	2.00000	0.00000	1.00000
1	1.00000	1.00000	1.00000	1.00000	-
2	0.50000	0.50000	0.49999	0.50000	-
3	0.16667	0.16667	0.16667	0.16667	-
4	0.41701(-1)	0.41683(-1)	0.41452(-1)	0.41667(-1)	-
5	0.82151(-2)	0.82686(-2)	0.89102(-2)	0.83333(-2)	-
6	0.15521(-2)	0.14883(-2)	0.78161(-3)	0.13889(-2)	-
7	0.80763(-4)	0.11956(-3)	0.51915(-3)	0.19841(-3)	-
8	0.72629(-4)	0.59917(-4)	-0.63478(-4)	0.24802(-4)	-
9	-0.80321(-5)	-0.58727(-5)	0.13956(-4)	0.27557(-5)	-
10	-0.14387(-5)	0.12896(-5)	-0.11451(-7)	0.27557(-6)	-
11	-0.10587(-7)	0.83539(-8)	0.11860(-5)	-	0.00000
12	-0.50000	-0.50000	-0.50000	-	-0.50000
13	-0.24926(-7)	0.47622(-7)	-0.31840(-5)	-	0.00000
14	0.41673(-1)	0.41670(-1)	0.41682(-1)	-	0.41667
15	-0.23763(-4)	-0.14315(-4)	-0.23127(-4)	-	0.00000
16	-0.13552(-2)	-0.13664(-2)	-0.13610(-2)	-	-0.13889(-2)
17	-0.25081(-4)	-0.18261(-4)	-0.21875(-4)	-	0.00000
18	0.35111(-4)	0.33169(-4)	0.34722(-4)	-	0.24802(-4)
19	-0.25048(-5)	-0.21233(-5)	-0.24527(-5)	-	0.00000
20	0.73300(-13)	-0.26372(-7)	0.39235(-9)	-	-0.27557(-6)

TABLE 6

(C) The quadrant. Finally, calculations were performed for the boundary value problem defined over a quadrant Q of the circular region of radius π, with center at the origin and lying in the first quadrant. As before, the boundary values were chosen so that the solution was $u(x,y) = e^x + \cos y$, and the approximating functions (10.12) were used. These calculations added a new feature in that certain integrals over Q and on parts of the boundary could not be evaluated exactly, contrary to the previous cases, but had to be computed numerically. For this purpose a 64-point Gaussian quadrature formula was used. Results for the standard case, N = 10, are

norm err. bound $= 6.60 \times 10^{-5}$

max. rel. err. $= 3.74 \times 10^{-5}$

computation time = 0.70 sec.

The coefficients of the trial functions for this case are the third column of a_i's of Table 6. The computation time given in this case is somewhat understated because it does not include the time needed to perform the numerical integrations involved.

CONCLUDING REMARKS

The central fact about the method is that the accuracy of the approximation is highly dependent on the choice of trial functions. Secondly, an analysis of the relative values of the coefficients of the trial functions and the components of the norm error can be very useful in indicating further trial functions to select from a given set to increase the accuracy of the approximation.

Some of the advantages of the method are the following:

(i) Since one is free to choose the trial functions, known information about the problem can be incorporated into the approximation.

(ii) The method is applicable over reasonably general regions. This includes regions for which it is usually difficult to apply finite difference and finite element methods or for which the accuracy of these methods is poor. However, applying the method over regions with irregular boundaries is usually more time-consuming and can lead to less accurate results since (a) integrals may have to be evaluated numerically, (b) eigenvalues which are needed in the determination of constants may not be known exactly for such regions, in which case a lower bound is required, and (c) the selection of suitable f^i may be more difficult.

(iii) Norm and pointwise bounds are available.

(iv) Computation time is very short, in general.

Some of the disadvantages are:

(i) For some problems a fair amount of analysis must be done prior to the actual computations. However, as one obtains more experience in the use of the method, the analysis time drops rapidly.

(ii) There are some problems in which the complexity of the integrals to be evaluated can impose limits on the freedom of choice of trial functions. We have found that there are times when the "natural" trial functions must be replaced by simpler functions in order to keep the analysis and programming within reasonable limits. This has been implicit in the examples we have given here in which the trial functions used lead to extremely simple integrals to evaluate.

References

1 G. Baronblat, I. Zheltov and I. Kochiva, Basic concepts in the theory of seepage of homogeneous liquids in fissured rocks, J. Appl. Math. Mech 24 (1960), 1286-1303.

2 J. H. Bramble and L. E. Payne, A priori bounds in the first boundary value problem in elasticity, J. Res. NBS 65B (1961), 269-276.

3 J. H. Bramble and L. E. Payne, Bounds in the Neumann problem for second order uniformly elliptic operators, Pacific J. Math. 12 (1962), 823-833.

4 J. H. Bramble and L. E. Payne, Bounds for solutions of second order elliptic partial differential equations, Contributions to Differential Equations 1 (1963), 95-127.

5 J. H. Bramble and L. E. Payne, Error bounds in the pointwise approximation of solutions of elastic plate problems, J. Res. NBS 57B (1963), 145-155.

6 J. H. Bramble and L. E. Payne, Pointwise bounds in the first biharmonic boundary value problem, J. Math. Phys. 42 (1963), 278-286.

7 J. H. Bramble and L. E. Payne, Some integral inequalities for uniformly elliptic operators, Contributions to Differential Equations 1 (1963), 129-135.

8 J. H. Bramble and L. E. Payne, Inequalities for solutions of mixed boundary value problems for elastic plates, J. Res. NBS 68B (1964), 75-92.

9 P. J. Chen and M. E. Gurtin, On a theory of heat conduction involving two temperatures, Z. Angew. Math. Phys. 19 (1968), 614-627.

10 G. Faber, Beweiss dass unter aller homogenen Membranen von gleicher Fläche und gleicher Spannung die Kreisförmige den tiefsten Grundton gibt, Sitz. Bayer. Akad. Wiss. (1923), 169-172.

11 G. Fichera, Su un principio di dualità per talune formole di maggiorazione relative alle equazioni differenziali, Atti Accad. Nazl. Lincei 19 (1955), 411-418.

12 B. Finlayson, The Method of Weighted Residuals and Variational Principles, Academic Press, N. Y. (1972).

13 D. W. Fox and W. C. Rheinbolt, Computational methods for determining lower bounds for eigenvalues of operators in Hilbert space, SIAM Rev. 8(1966), 427-462.

14 L. Fox, P. Henrici and C. Moler, Approximations and bounds for eigenvalues of elliptic operators, SIAM J. Num. Anal. 4(1967), 89-103.

15 A. Friedman, Partial Differential Equations of Parabolic Type, Prentice-Hall, Englewood Cliffs, New Jersey (1964).

16 F. John, Plane Waves and Spherical Means Applied to Partial Differential Equations, Interscience Tracts No. 2, New York (1955).

17 J. R. Kuttler and V. G. Sigillito, Optimal constants in _a priori_ inequalities, unpublished manuscript.

18 J. R. Kuttler and V. G. Sigillito, Inequalities for membrane and Stekloff eigenvalues, J. Math. Anal. and Appl. 23 (1968), 148-160.

19 J. R. Kuttler and V. G. Sigillito, Explicit L_2 inequalities for parabolic and pseudoparabolic equations with Neumann boundary conditions, unpublished manuscript.

20 J. R. Kuttler and V. G. Sigillito, Bounding eigenvalues of elliptic operators, SIAM J. Math. Anal. (to appear).

21 E. A. Milne, The diffusion of imprisoned radiation through a gas, J. London Math. Soc. 1(1926), 40-51.

22 C. Miranda, Partial Differential Equations of Elliptic Type, Springer-Verlag, Berlin (1970).

23 A. C. G. Mitchell and N. W. Zemansky, Resonance Radiation and Excited Atoms, Cambridge University Press, Cambridge, England (1934).

24 C. B. Moler and L. E. Payne, Bounds for eigenvalues and eigenvectors of symmetric operators, SIAM J. Num. Anal. 5(1968), 64-70.

25 K. L. E. Nickel, Extension of a recent paper by Fox, Henrici and Moler on eigenvalues of elliptic operators, SIAM J. Numer. Anal. 4(1967), 453-488.

26 V. O'Brien, Pulsatile fully developed flow in rectangular channels, J. Franklin Inst. 300(1975), 225-230.

27 L. E. Payne, Inequalities for eigenvalues of membranes and plates, J. Rational Mech. Anal. 4(1955), 517-528.

28 L. E. Payne and H. F. Weinberger, New bounds in harmonic and biharmonic problems, J. Math. Phys. 4(1955), 291-307.

29 L. E. Payne and H. F. Weinberger, Lower bounds for vibration frequencies of elastically supported membranes and plates, J. Soc. Ind. Appl. Math. 5(1957), 171-182.

30 L. E. Payne and H. F. Weinberger, New bounds for solutions of second order elliptic partial differential equations, Pacific J. Math. 8(1958), 551-573.

31 L. E. Payne and H. F. Weinberger, A Faber-Krahn inequality for wedge-like domains, J. Math. and Phys. 39 (1960), 182-188.

32 L. E. Payne and H. F. Weinberger, An optimal Poincare inequality for convex domains, Arch. Rational Mech. Anal. 5(1960), 286-292.

33 L. E. Payne, Isoperimetric inequalities and their applications, SIAM Review 9 (1967), 453-488.

34 J. C. Pirkle and V. G. Sigillito, A priori inequalities and norm error bounds for solutions of a third order diffusion-like equation, SIAM J. Appl. Math. 25(1973), 69-71.

35 J. C. Pirkle and V. G. Sigillito, Analysis of optically pumped CO_2 laser, Applied Optics 13(1974), 2799-2807.

36 G. Pólya and G. Szegö, Isoperimetric Inequalities in Mathematical Physics, Ann. of Math. Studies No. 27, Princeton University Press, Princeton (1951).

37 F. Rellich, Darstellung der Eigenwerte von $\Delta u + \lambda u$ durch ein Randintegral, Math. Z. 46 (1940), 635-646.

38 V. G. Sigillito, A priori inequalities and interior pointwise bounds for solutions of certain parabolic and elliptic partial differential equations, Doctoral Dissertation, U. of Md., 1965.

39 V. G. Sigillito, Pointwise bounds for solutions of the first initial-boundary value problem for parabolic equations, J. SIAM Appl. Math. 14(1966), 1038-1056.

40 V. G. Sigillito, On a continuous method of approximating solutions of the heat equation, J. Assoc. Comput. Mach. 14(1967), 732-741.

41 V. G. Sigillito, Pointwise bounds for solutions of semilinear parabolic equations, SIAM Review 9 (1967), 581-585.

42 V. G. Sigillito, A priori inequalities and pointwise bounds for solutions of fourth order elliptic partial differential equations, SIAM J. Appl. Math. 15(1967), 1136-1155.

43 V. G. Sigillito, Exponential decay of functionals of solutions of a pseudoparabolic equation, SIAM J. Math. Anal. 5(1974), 581-585.

44 V. G. Sigillito, A priori inequalities and the Dirichlet problem for a pseudoparabolic equation, SIAM J. Math. Anal. 7(1976), 222-229.

45 V. G. Sigillito, A priori inequalities and approximate solutions of the first boundary value problem for $\Delta^2 u=f$, SIAM J. Numer. Anal. 13(1976), 251-260.

46 V. V. Sobolev, A Treatise on Radiative Transfer, Van Nostrand, New York (1963).

47 J. T. Stadter, Bounds to eigenvalues of rhombical membranes, SIAM J. Appl. Math. 14(1966), 324-341.

48 T. W. Ting, Certain non-steady flows of second-order fluids, Arch. Rational Mech. Anal. 14(1963), 1-26.

SUPPLEMENTARY REFERENCES: NOT CITED IN THE TEXT BUT OF INTEREST

S1 F. Bellar, Pointwise bounds for the second initial-boundary value problem of parabolic type, Pacific J. Math. 19(1966), 205-219.

S2 J. H. Bramble and L. E. Payne, Bounds for derivatives in the Dirichlet problem for Poisson's equation, J. Soc. Ind. Appl. Math. 10(1962), 370-380.

S3 J. H. Bramble and L. E. Payne, Upper and lower bounds in equations of forced vibration type, J. Rational Mech. Anal. 14(1963), 153-170.

S4 J. H. Bramble and L. E. Payne, Bounds for derivatives in elliptic boundary value problems, Pacific J. Math. 14(1964), 777-782.

S5 A. Elcrat, Constructive existence for semilinear elliptic equations with discontinuous coefficients, SIAM J. Math. Anal. 5(1974), 663-672.

S6 A. Elcrat and V. G. Sigillito, An explicit a priori estimate for parabolic equations with applications to semilinear equations, SIAM J. Math. Anal. 7(1976), 746-753.

S7 A. Elcrat and V. G. Sigillito, Coercivity for a third order pseudoparabolic operator with applications to semilinear equations, J. Math. Anal. Appl. (to appear).

S8 K. Gustafson, Stability inequalities for semimonotonically perturbed nonhomogeneous boundary problems, SIAM J. Appl Math. 15(1967), 368-391.

S9 K. Gustafson and V. G. Sigillito, Inequalities for nonlocal parabolic and higher order elliptic equations, SIAM Rev. 9(1967), 351-541.

S10 L. E. Payne, Error bounds based on a priori estimates, in Numerical Solution of Partial Differential Equations, ed. by J. H. Bramble Academic Press, N. Y. (1966).

S11 L. E. Payne and H. F. Weinbeiger, Bounds for solutions of second order elliptic equations in terms of arbitrary vector fields, J. Rational Mech. Anal. 20(1965), 95-106.